Infrared Spectroscopy of Diatomics for Space Observation

Infrared Spectroscopy Set

coordinated by
Pierre-Richard Dahoo and Azzedine Lakhlifi

Volume 1

Infrared Spectroscopy of Diatomics for Space Observation

Pierre-Richard Dahoo
Azzedine Lakhlifi

iSTE

WILEY

First published 2017 in Great Britain and the United States by ISTE Ltd and John Wiley & Sons, Inc.

ISTE Ltd
27-37 St George's Road
London SW19 4EU
UK

www.iste.co.uk

John Wiley & Sons, Inc.
111 River Street
Hoboken, NJ 07030
USA

www.wiley.com

Library of Congress Control Number: 2017953145

British Library Cataloguing-in-Publication Data
A CIP record for this book is available from the British Library
ISBN 978-1-78630-116-1

MIX
Paper from responsible sources
FSC
www.fsc.org
FSC® C013604

Contents

Foreword

Light is one of the vital elements for the living world. Our sun provides us with not only energy but also light, the latter of which reveals the beauty of that which surrounds us, and reveals to us shapes, textures, colors, shadows, reflections, and more. Light also reveals thousands of nuances that enable us to perceive the world around us and that make sight our finest, most developed and most useful sense in everyday life. Even if we are less frightened by it than our ancestors were, the luminous phenomena surrounding us are still as surprising and intriguing. They often seem inexplicable and complex as they are multiple, variable, changing, fast and impalpable.

However, none of this proceeds from magic and can be explained by physical phenomena, most of which have already been interpreted and modeled. The aim of spectrometry is to strive to understand the interactions between light and matter and, by analyzing changes in the color of light before and after interactions, to identify the interacting environment and in particular its composition.

In the atmosphere, an environment characterized by its low density, light allows this analysis to be made from a distance. This particularity has allowed us to improve our knowledge of the environments that surround us; from rainbows caused by drops of water, to the great northern lights, to comet tails, and all the way to distant galaxies. This information enables us to remove ourselves from Earth's attractions and to explore these worlds. The number of exoplanets being discovered keeps on increasing, while a little over 20 years ago, their existence was just speculation and the subject of science fiction.

But the eye is also a limited detector and particularly so for the infrared range. Yet this range of frequency has its own assets in both our everyday life as well as for scientists. It enables us to switch on the television with a remote control, but also, for more technical applications, to analyze numerous chemical species and to give information on the speed of objects, their temperature and many other parameters.

Today, this field of application is far from being an exhausted and perfectly mastered science, but some fabulous progress has been made over the last decade. This knowledge and simultaneously the possibilities opened for space exploration has permitted some considerable scientific advancement. Harnessing these advances has allowed us today to further our understanding of our atmosphere and our exploration of the worlds surrounding us, even those we cannot see, or cannot imagine. This knowledge is essential for the modern explorers that we are, whose utmost aim is directed toward improving the description of these worlds thanks to teledetection.

Capitalizing and gathering this knowledge is therefore an endeavor that is necessary for pursuing our quest for knowledge. Beware, delving into these phenomena may cause vertigo. Physics and chemistry combine on the molecular, atomic, electronic, photonic, etc. scales. Mastering them will enable you to understand the fundamentals of physics, and to develop more and more sophisticated instruments and methods to extract multiple useful pieces of information from radiation. This information will turn you into the sorcerers of tomorrow, capable of decoding physical–chemical processes via light.

Philippe KECKHUT
Versailles Saint-Quentin-en-Yvelines University
and Director of LATMOS

Preface

Theoretical and instrumental spectroscopies optimize the parameters that enable chemical species, molecules, radicals and ions to be characterized based on theoretical models and experimental data. The results of research undertaken in laboratories are used to analyze the data provided by different instruments used in astronomy for observing the sky or space and detecting atoms, molecules or chemical species (ions, radicals, macromolecules, nanocages, etc.) located in different environments ranging from the atmosphere of planets, including Earth, and their satellites to interstellar media, comets and exo-planets.

Continuous progress within the field of the technology of embedded and communicating systems enables a mass of data to be retrieved with greater and greater precision (high-resolution spectra, weak signal-to-noise ratio of more and more effective detectors, more reliable transmission and data-retrieval techniques) from instruments implemented in observatories based on the ground or in space probes and space-borne telescopes. Thanks to sky-viewing telescopes or telescopes coupled with measuring instruments, spectrographs, spectrometers and interferometers equipped with standard detection systems or more sophisticated ones, using, for instance, intensity autocorrelation or heterodyne detection techniques, many discoveries and advances in astrophysics have been made possible. For example, without being too exhaustive, one can list some of the following advances from space observation:

– Galileo Galilei, who was one of the first astronomers to have used a telescope (a word he invented in 1611) for observing the sky and the stars. In his writing concerning his observations, his discovery in 1610 that four moons revolve around Jupiter led him to adopt the heliocentric model of the universe, as proposed by Copernicus in 1543 against the geocentric model, which was taken for granted under the authority of the scientific community and the church;

– in an article in 1919, F.O. Dyson, A.S. Eddington and C. Davidson used Eddington's photographs taken during a solar eclipse as the basis for presenting the determination of a star's light deviation by the Sun's gravitational field. These findings then confirmed the gravitational field as predicted by Einstein's theory of relativity;

– later, in 1929, the expansion of the Universe was discovered by Edwin Hubble, on the basis of the redshift (Doppler effect) of the galaxies from the spectrum of Cepheids and stars analysis characterized by variations in brightness, observed in the Hooker telescope of the Mount Wilson Observatory in California;

– by improving observation apparatuses, in 1956, Handbury Brown and Twiss used the autocorrelation technique with an intensity interferometer to measure the angular diameter of the Sirius star in the optical range;

– in 1965, fossil radiation or cosmic microwave background radiation (CMBR), predicted by Ralph Alpher in his thesis in 1948, was discovered by Arno Penzias and Robert Wilson at the Bell Telephone Laboratories when developing a new radio receiver after interferences were detected, regardless of the orientation of the antenna being constructed and corresponding to 3.5 K in temperature for a wave length of 7.35 cm;

– in 1976, M.A. Johnson, A.L. Betz, R.A. McLaren, E.C. Sutton and C.H. Townes observed a mesospheric emission in a non-local thermodynamic equilibrium system ("non-LTE") coming from Mars as well as Venus. This phenomenon, which was discovered for the first time, was later interpreted as a CO_2 stimulated emission in the infrared region around 10 μm by D. Deming and M.J. Mumma in 1983;

– in 1990, J.C. Mather *et al.* determined the characteristic CMBR temperature value to be at 2.735 K based on data transmitted by COBE ("cosmic background explorer") satellite launched in 1989, equipped with FIRAS ("far infrared absolute spectrometer") an instrument that measured the fossil radiation spectra of the cosmic microwave background more precisely;

– in 1995, a planet around a solar-type star, 51 Peg b was observed for the first time by Mayor and Quelozà of the Hautes Provence Observatory (OHP), using the Doppler effect measurement of the spectrum obtained with the Elodie spectrograph following the periodic variations of the radial speed of the host star at 56 m/s in 4 days;

– using a new type of vacuum cross-dispersion spectrometer, in 1999 P. Connes and F. Bouchy discovered, again at OHP, another exo-planet around the star Upsilon Andromedae with a measurement accuracy of 3 m/s;

– and very recently, on September 14, 2015, LIGO ("Laser Interferometer Gravitational-Wave Observatory"), detected distortions caused by gravitational waves in space time for the first time. This phenomenon which was predicted by Einstein's theory of general relativity, was generated by two black holes that collided at about 1.3 billion light years.

A special mention should be made about Rosetta, the spacecraft on the ESA mission, with contributions from the EU Member States and NASA. It was placed into orbit around the comet of the Jupiter 67P/Churyumov-Gerasimenko (67P) of the Jupiter family. Comets are the best sample of primitive solar nebulous material currently available and which date back 4.57 billion years to the origin of our planetary system. Rosetta found the comet's nucleus on August 6, 2014, at 3.7 astronomical units (AU) from the Sun, and landed the Philae spacecraft on the surface of the nucleus on November 12, 2014, when the comet was at 3.0 AU from the Sun.

The initial observations enabled:

– a reference frame to be made that described the global shape, morphology and composition of the surface, as well as the global physical properties of the nucleus;

– various molecules to be detected, including $H_2{}^{17}O$, $H_2{}^{18}O$, CO and CO_2, and for their temporal variability, as well as their heterogeneous distribution to be evaluated;

– the measurement of the high D/H ratio in water, 5.3×10^{-4}, which excludes the hypothesis that comets from the Jupiter family only contain ocean water from Earth.

The Rosetta mission started exploring our origins because of the efforts of thousands of people at ESA, at NASA, in the industry, in space agencies

Amat, Nielsen and Tarrago, Barchewitz, Hollas, Lefebvre-Brion and Field, Brown, Landau, Cohen-Tannoudji or Messiah. The list would be too long and mention is only made of those who were consulted for this part, to which we should add G. Amat's Molecular Physics course at Master 2 level at UPMC and the courses of J.M. Flaud and C. Camy-Peyret in second year master's (Master 2 level) degree in DEA (Diplôme d'Études Approfondies) Laser and Matter at UPSUD. The second part, which also comprises two chapters, describes the theoretical models developed in the research field and applied to the analysis of experimental results in the CNRS laboratories, a national research center in France. These works were initiated particularly in the Molecular Physics group of Besançon (L. Galatry, D. Robert, J. Bonamy, L. Bonamy, C. Girardet, A. Lahklifi, etc.) and later collaboratively pursued with laboratory researchers in the region of Paris (C. Boulet, J.M. Hartmann, P.R. Dahoo, etc.) to study molecules in different media and subjected to interactions whose effects, particularly on the nanometric scale, modify the IR spectral signature of the molecules. Finally, in the third part, we introduce some applications of the models described in the second part to study some diatomic molecules.

Chapter 1 deals with diatomic molecules in general, giving the full Hamiltonian operator, which enables us to determine their different electronic states and, within the framework of Born and Oppenheimer approximation, their vibration–rotation states. The symmetry properties of homonuclear and heteronuclear molecules, as well as the nomenclature of the different electronic levels of vibration–rotation degrees of freedom based on the corresponding symmetry groups, are also recalled. Chapter 2 introduces the models developed for studying, in the near and mid infrared ranges, the vibration spectroscopy of a diatomic molecule in its fundamental electronic state. Also described are the Rayleigh–Schrödinger's standard perturbation theory and the non-standard contact perturbation theory proposed by Van Vleck for determining the states of vibration when one considers the non-harmonicity of the electronic potential driving the movement of the nuclei. Theories for calculating the states of rotation in the far IR range are also addressed, as well as the effect of coupling between the vibrational and rotational degrees of freedom on the corresponding spectra. Chapters 3 and 4 remind the reader of the theories developed for analyzing the spectra of diatomic molecules when they are in a constrained environment, under the effect of pressure in gas phase or in nanocages in condensed media. The Liouville theorem in the Liouville line space and the linked cluster theorem through the cumulant expansion, particularly developed in the field of statistical physics, are applied to study the effect of interactions between the active molecule and its environment in a system characterized by large degrees of freedom. The shifts and widths of lines are calculated as well as the modifications of the spectra, particularly those connected to the rotation in the condensed phase. In Chapter 5, the theories described in previous chapters are applied to different types of diatomic molecules to illustrate how the different applications allow the spectra observed in Raman spectroscopy with a lidar, in classical absorption spectroscopy, to be analyzed and interpreted when the molecule is in gas phase under pressure, in a nanocage, or on a surface.

Pierre-Richard DAHOO
Azzedine LAKHLIFI
September 2017

1

Generalities on Diatomic Molecules

A diatomic molecule is the simplest possible assembly based on a system of atoms that lead to molecules, whether homonuclear or heteronuclear. Its spectral signature is the result of the interaction of an electromagnetic radiation with the electrons, given the movement of its charged constituents, nuclei and electrons. This interaction is modeled by an operator, the n-polar moment, which leads to a transition between the diatomic molecule's energy states, the eigenfunctions of the Hamiltonian, its energy operator built on the classical (electronic, vibration and rotation) and quantum (electronic and nuclear spin) degrees of freedom. The resolution of the Schrödinger eigenvalues equation of the molecular system is based on the Born–Oppenheimer (BO) approximation, which allows the electrons' fast movement to be decoupled from that of the nuclei, which is much slower, and to separately process its electronic and vibration–rotation degrees of freedom in spectroscopy. The characteristics of the electronic, vibrational and rotational spectra depend on the molecule's symmetries and its electronic and nuclear spin properties. We can use group theory for predicting the absorption spectra profiles and the transition rules expected in infrared or Raman spectroscopy. Thus, molecular species are divided into para and ortho varieties for the H_2 molecule, for instance. In the Raman rotational spectrum, H_2 presents an intensity alternation which is different to that of N_2 as a function of the rotational quantum number, whereas for O_2, one line out of two is absent in the spectrum.

1.1. Generalities on detecting diatomic molecules

1.1.1. *Radiation–matter interaction for detection*

The presence of a molecule may be revealed by probing a given medium with an electromagnetic radiation favorable for it to be observed in the frequency range corresponding to radio waves, microwaves, millimetric waves, Térahertz radiation, infrared radiations (far, mid, near), visible light, ultraviolet radiations or X-rays. Different techniques have been developed for studying molecules according to the type of interaction (linear, nonlinear, emission, absorption, diffusion, etc.) and the spectral region.

As a general rule, the interaction of an electromagnetic radiation with a molecule (Figure 1.1), whether diatomic or polyatomic, results in physical–chemical processes such as the absorption or emission of photons (spectroscopy of absorption or emission), elastic diffusion (Rayleigh, Mie) or inelastic diffusion (Raman, Brillouin), phenomena of dissociation, ionization or a combination of these processes. These processes involve transitions between discrete energy levels and a continuum of levels following the interaction between the multipolar moments of the molecule and the electrical field of the radiation. The interaction between an electromagnetic wave and matter was discussed more thoroughly in a previous publication (Chapter 3 [DAH 16]).

To each state of a molecule, the quantum mechanics theory associates a wave function that depends on the position coordinates and the spins of the two nuclei and electrons comprising the molecule. In order to determine the molecule's quantum state, one must solve the Schrödinger eigenvalues equation of the molecular system:

$$H\left|\Psi\right\rangle = E\left|\Psi\right\rangle, \qquad [1.1]$$

where H is the energy operator, corresponding to the Hamiltonian of the molecular system and $\left|\Psi\right\rangle$ is the wave function of the system expressed in Dirac's notation.

In the case of an electric dipolar transition between the states $\left|\Psi_i\right\rangle$ and $\left|\Psi_f\right\rangle$, the probability of a radiative transition is given by Einstein's spontaneous emission coefficient:

$$A_{if} = \tau^{-1} = \frac{64\pi^4}{h}\frac{g_i}{g_f}\nu_{if}^3\left|R_{if}\right|^2 \qquad [1.2]$$

where τ is the radiative lifetime, g_i and g_f are the degeneracies of the levels i and f, ν_{if} is the frequency of the transition considered and R_{if} is the matrix element of the dipolar moment of the transition. It can be shown that the Einstein spontaneous emission coefficient is connected to the induced absorption coefficient B_{if} or to the induced emission coefficient B_{fi} by the relation:

$$A_{if} = \frac{8\pi h\nu_{if}^3}{c^3}B_{if} = \frac{8\pi h}{\lambda_{if}^3}B_{if} = 8\pi h\sigma_{if}^3 B_{if} \qquad [1.3]$$

with $\frac{B_{if}}{g_f} = \frac{B_{fi}}{g_i}$ and in which λ_{if} is the wave length and σ_{if} (in cm^{-1}) the wave number associated with the transition between the two levels.

Based on the balanced equation between the levels involved in the transition and by considering the populations of each level such that: $N_i = N_0\, exp\left(-\frac{E_i}{k_B T}\right) = N_0\, exp\left(-\frac{hc\sigma_i}{k_B T}\right)$, as well as the shape of the absorption line (Gaussian by Doppler effect, Lorentz by pressure broadening), the intensity of an absorption vibration–rotation line in gas phase in the infrared range can be written as:

$$S_{if}^g = \int_{\sigma_{\min}}^{\sigma_{\max}} \frac{8\pi^3}{3hc}\sigma_{if} f\left(\sigma-\sigma_{if}\right) N_0 \left[1 - e^{\left(-\frac{hc\sigma_{if}}{k_B T}\right)}\right] g_i \frac{e^{\left(-\frac{hc\sigma_i}{k_B T}\right)}}{Q(T)} R_{if}^2\, d\sigma \qquad [1.4]$$

where σ_{if} is the wave number (in cm^{-1}) of the rotational–vibrational transition considered, N_0 is the number of molecules per volume unit, $hc\sigma_i$ is the energy of the initial level, $Q(T) = \sum_i g_i \exp\left(-E_i/k_B T\right)$ designates the total partition function of the molecule at the temperature T, R_{if} is the rotational–vibrational transition moment and finally g_i is the degeneracy due to the nuclear spin of the initial level. In this formula, the line shape is a Voigt function, the convolution product of a Gaussian by a Lorentzian, because the Doppler effect and the broadening by pressure are concomitant processes and its integral on the absorption domain equals 1 ($\int_{\sigma_{\min}}^{\sigma_{\max}} f(\sigma-\sigma_{if})\,d\sigma = 1$).

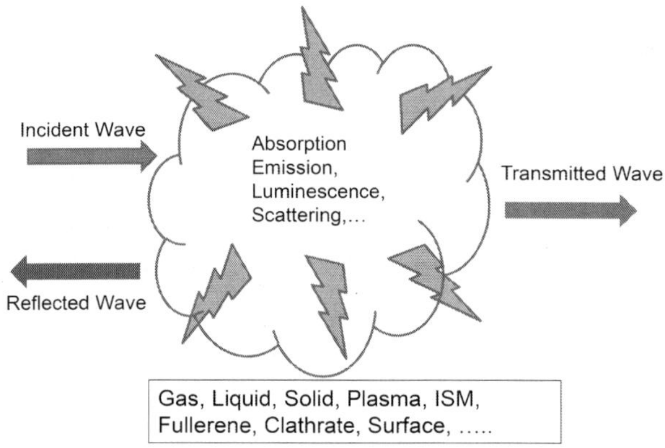

Figure 1.1. *Light–matter interaction*

When the absorption lines corresponding to the transitions between the energy levels can be experimentally determined, either by classical spectroscopy resolved in frequency via an interferometer or by laser absorption by sweeping the absorption spectral domain of the line, one can connect this intensity to the absorption coefficient as determined by the Beer–Lambert law.

The absorption coefficient $\alpha(\omega)$ of a gas probed on a length dz by an electromagnetic wave at the angular frequency ω is expressed as $\alpha(\omega) = \frac{1}{I}\frac{dI}{dz}$, where I is the temporal average of the Poynting vector of the electromagnetic radiation or intensity ($I = \frac{cn}{2}|E_0|^2$). In QM, its expression is given by: $\alpha(\omega) = \frac{\hbar\omega\Gamma}{PV}$, where P is the Poynting vector representing the flux of the incident electromagnetic wave, V is the volume, Γ is the transition probability per unit time and $\hbar\omega$ is the energy of the absorbed photon. For a transition in gaseous phase, the absorption coefficient $\alpha(\omega)$ as a function of the angular frequency ω depends on the molecule, the transition, the number of molecules, the pressure and the temperature. Experimentally, we can extract these quantities from the integrated intensity of the absorption spectra:

$$S_{if}^{g} = \frac{1}{2lN_0}\int_{\sigma_{\min}}^{\sigma_{\max}} \ln\frac{I_0\left(\sigma-\sigma_{if}\right)}{I_t\left(\sigma-\sigma_{if}\right)}d\sigma = \frac{1}{2lN_0}I_{if} \qquad [1.5]$$

where N_0 is the number of molecules per cm^3, $2l$ is the length traveled by the radiation in the sample and the term contained in the integral is the integrated absorption measured on the IR absorption spectrum domain. Another parameter that enables the absorption to be quantified is the integrated absorption coefficient $S_{if}^{0} = \frac{S_{if}}{P}$, which is expressed relative to the pressure of the surrounding media.

From the observed isolated line profiles that are connected to the molecular transitions of the species present in the planetary atmospheres (*atmos:* vapor; *sphaira:* sphere), it is possible to extract physical parameters such as the pressure, temperature, molecular abundance and the dynamic properties of the atmosphere, such as the wind speed and the circulating currents associated with the movement of chemical species from hot regions to cold ones, for example.

1.1.2. *Diatomic molecules: observation, analysis and interpretation*

Studies based on the spectra and the photographs obtained from observations of the sky, either performed in ground-based observatories or in various space missions, have validated the existence of an atmospheric layer on other planets of the solar system, just like on Earth. For example, in a diatomic molecular form, amongst other major constituents of the planets' atmospheres, one finds dinitrogen N_2 and dioxygen O_2 on Earth and dihydrogen H_2 on Jupiter. On Triton and Titan, the largest satellites of the planets Neptune and Saturn, respectively, the atmosphere is mainly made up of dinitrogen N_2 similar to Earth's atmosphere. The identification of these diatomic species has enabled us to compare the planets' composition and to propose scenarios on the environmental conditions that prevailed when the planets were formed. Giant planets were formed in an environment that essentially contains hydrogen (75% in mass) and helium (24%), a cloud pushed away by solar wind up to the distance of Jupiter and close in composition to the primitive nebula at the origin of the Sun. As for the four telluric planets, they were formed in very similar conditions, based on the same materials, but with a gaseous environment different in composition to that of the primitive nebula.

The negligible quantity of dihydrogen H_2 on telluric planets (Mars, Earth and Venus) in comparison to the Jovian planets (Jupiter and Saturn) is explained by a stronger force of gravity and weaker temperatures of the external layers compared to the telluric planets. Therefore, the external Jovian planets preserved an atmosphere which is close in composition to that of the primitive nebula.

The singularity of the 20% of O_2 in the Earth's atmosphere is attributed to two processes. A physical–chemical process takes place in the high Earth atmosphere by which solar and cosmic radiation breaks down water molecules in dioxygen and dihydrogen. As the dihydrogen definitively escapes into space at an average rate of 3 kg/s, the dioxygen accumulates in the atmosphere. By a biochemical natural process, O_2 is also produced by cyanobacteria, and more generally by green plants that transform carbon dioxide into dioxygen by the mechanism of photosynthesis.

Nitric oxide (NO) emissions can be observed on all telluric planets surrounded by an atmosphere. On Venus, these emissions have been identified with those of O_2 in the high atmosphere of Venus since 1979 [CON 79, FEL 79, STE 79]. These two molecules are detectable because of their aeronomic emissions on the nocturnal hemisphere of the planet. Dioxygen O_2 mainly emits in the infrared range and NO in the ultraviolet range between 180 and 300 nm. It is also possible to detect NO in the near infrared between 1.2 and 1.3 μm. The aeronomic emissions of NO are due to a radiative recombination process, which is produced on the night side of the planet. As Venus's atmosphere is essentially composed of CO_2 and N_2, solar radiation breaks these molecules into free atoms of nitrogen and oxygen on the day side. In the high atmosphere, and above 100 km, the zonal winds transport these atoms toward the nocturnal side where they recombine to form NO in an excited state, which then emits ultraviolet radiation. The aeronomic emission of NO is thus a natural tracer of subsolar/antisolar circulation.

On Mars, the observations from the imaging spectrometer OMEGA on ESA's Mars Express orbiter led to the detection of a dim infrared light above Mars' Winter poles. There again, when the oxygen and nitrogen atoms recombine, an ultraviolet nocturnal light results from the NO formed, as well as an emission in the near infrared (O_2) at 1.27 μm from dioxygen. The detection of the nocturnal telluric emission was observed for the first time in 2010 but it was expected ever since 2005 when the NO emission in Mars' atmosphere was detected by the SPICAM instrument on board Mars Express. The Mars nocturnal fluorescence of O_2 ($a^1\Delta_g$) at 1.27 μm thus detected is due to the recombination of the O atoms formed by the photolysis of CO_2 during the day at higher altitudes at 80 km and transported toward the bottom at the pole by the Hadley circulation. In 2010, the first detections of the nocturnal fluorescence of O_2 ($a^1\Delta_g$) indicated that it is about two orders of magnitude less intense than the daytime fluorescence that has a different source, being due to the photodissociation of the ozone molecule. Similar observations have been made from the Venus Express mission on the fluorescence of the atmosphere of Venus in good correlation with the observations made from the ground base observatories for this planet [BER 12, GER 08, GER 09a, GER 09b, GON 10].

The detection of gaseous matter between some billions of suns of the Milky Way in space has led the scientific community in astrophysics to postulate the existence of an interstellar medium (ISM), which is composed of dense or diffuse molecular clouds or dust grains. In the ISM, which in a laboratory would be considered to be a hypervacuum environment [TEN 03], the gaseous matter exists in different forms of rather cold clouds at temperatures of the order of 20–30 K. The interstellar molecules are formed by chemical reactions in interstellar clouds of gas and interstellar and circumstellar dust. When a molecule is ionized as a result of its interaction with cosmic radiation, the electrically charged ion tends to attract nearby reactants through electrostatic interaction on the dust surfaces.

Among the most abundant diatomic molecules detected one finds dihydrogen H_2 and in molecular clouds, carbon monoxide CO whose presence is revealed by its rotational transition at 2.6 mm. The presence of dihydrogen is expected because the universe is mainly made up of atomic hydrogen and although predicted by A. Eddington as early as in 1926, it was only detected half a century later in 1970 [CAR 70]. In molecular clouds, almost all the hydrogen is under a molecular form. The Copernicus satellite has detected absorption bands of ultra-UV lines of H_2 (Lyman & Werner electronic transition bands) and HD at the beginning of the 1970s.

The first detections of diatomic species in the space surrounding the stars of our galaxy, CH, CH^+ and CN (cyanide embryo), which are very simple molecular species, were made between 1937 and 1941 [ADA 41, MCK 40, SWI 37] in the field of astrophysics, through observations of characteristic absorption lines present in the radiation spectrum of many stars in the ultraviolet and visible ranges. Given the electronic structure of these interstellar species, namely two radicals with dangling valences and a cation, it was difficult to consider them as ordinary chemical products. The presence of CH^+ was confirmed by laboratory experiments, which enabled the cation to be characterized at around 400 nm [DOU 41]. By 1941, diatomic molecules or species had been identified not only in solar and star spectra but also in those of comets, particularly in those observed in the ISM such as OH, C_2, NH, CO^+ and N_2^+, for example.

The analysis of the spectrum attributed to CN [MCK 40] showed structures with initial transition states not only from the fundamental rotational level but also from the first excited level. Moreover, the relative intensities of these lines were consistent with a population distribution of the two levels characterized by a temperature of 2.3 K. However, the cosmological importance of this temperature was not then acknowledged. Thanks to progress in technology that has helped to broaden the electromagnetic observation range of chemical species to include microwave, millimetric and infrared radiation, in 1965, Penzias and Wilson [PEN 65] observed in the microwave range an isotropic radiation that was attributed to the isotropic radiation of a black body at 3.5 K by Dicke *et al.* [DIC 65] (cosmological background radiation (CMBR) currently measured at 2.7 K). The development of radio-telescopes further enabled the detection of molecules, which have a permanent electric dipole moment. In 1960, the spectral absorption lines of the OH radical were observed at 18 cm (OH/H ratio of 10^{-7}) by Weinreib *et al.* [WEI 63] in the Cassiopeia A constellation. In 1970, the CO emission line, $J = 1 \rightarrow J = 0$ at 2.6 mm was detected for the first time by R.W. Wilson, K.B. Jefferts and A.A. Penzias. These diatomic molecules constitute molecular tracers with a wave length in the mm range, which enable probing of the physical conditions in the interstellar clouds (temperature, density of the molecular hydrogen, mass, velocity profiles, etc.) as shown by the pioneering works of Encrenaz *et al.* [ENC 75], Tucker *et al.* [TUC 76] and Dickman [DIC 78].

The technology of receivers and antennas has been continuously improving since the 1980s and numerous complex interstellar molecules have thus been detected, mainly with wavelengths in the millimeter range (the largest molecules are chains containing 13 atoms). The study of the physical–chemical processes in the ISM is paramount to fully understand and correctly interpret the observations from dense and cold molecular clouds, not only in our own galaxy but also in other galaxies, particularly in the case of CO studies. As mentioned in the Preface, progress in technologies enables us to design and build observation instruments (telescopes, LIDAR, spectrometers, interferometers, etc.) on board static or mobile observation platforms in order to probe the Earth's atmosphere and also the universe and follow the evolution of galaxies (NASA's James Webb space telescope for 2018) or even observe the exo-planets, the first days of the universe, supermassive black holes and the mysterious nature of dark matter and dark energy ("European Extremely Large Telescope", for 2024). All these instruments equipped with sensors and other detection systems are designed to match the relevant electromagnetic radiation necessary to investigate galaxies and the universe nearest to regions in space that will be observed and explored. Technologies of connected objects associated with drones and robotics will allow us to increase our observation capacities and data analysis must be supported by theoretical models of non-isolated molecules built on physical grounds and from first principles for relevant and objective interpretation.

1.2. Hamiltonian of a diatomic molecule

Nuclei and electrons are characterized by their masses (m_N and m_e), their charges ($Z_N e$ et $-e$), their spins (I_N and ½), their magnetic moments (μ_N and μ_e) and for nuclei, a quadrupolar moment ($Q^N_{\alpha\beta}$). Given these characteristics, when we consider the different interactions between the nuclei and the electrons, the Hamiltonian of the molecular system can be put in the following form:

$$H = T + V + H_{SF} + H_{shf} \quad [1.6a]$$

in which

$$T = T_N + T_e = T_{CM} + T_0 + T' \quad [1.6b]$$

$$V = V_{NN} + V_{ee} + V_{eN} \quad [1.6c]$$

$$H_{FS} = H_{SO} + H_{SR} + H_{SS} \quad [1.6d]$$

$$H_{hfs} = H_{IL} + H_{IR} + H_{II} + H_{IS} + H_{EQ} \quad [1.6e]$$

In equation [1.6b], $T_N = \sum_{N=1}^{2} \frac{P_N^2}{2M_N}$ is the kinetic energy operator of the nuclei and $T_e = \sum_{e=1}^{n} \frac{P_e^2}{2m_e}$ is the kinetic energy operator of the electrons. It can be shown that the sum of T_N and T_e can be expressed as the sum of T_{CM}, the kinetic energy of the center of mass (CM), T_0, the kinetic energy of particles inside the molecule and the mass polarization operator, a crossed kinetic term of the form $T' = \frac{1}{2m_N} \sum_{e,e'=1}^{n} P_e P_{e'}$, where P_e is the linear momentum operator of an electron that represents the fluctuations of the position of the CM due to the movement of the electrons and where $m_N = \frac{M_1 M_2}{M_1 + M_2}$ is the reduced mass of the nuclei. A very good discussion of this decomposition can be found in [BRO 03].

Equation [1.6c] represents the Coulomb potential that characterizes the electrostatic interaction between the nuclei (V_{NN}), between the electrons (V_{ee}) and between the electrons and the nuclei (V_{eN}).

Equation [1.6d] (H_{FS}, *FS*: fine structure) describes the interaction of the magnetic moments of the electrons with the intramolecular field that, in the case of a diatomic molecule, is of cylindrical symmetry. The first term H_{SO} (spin-orbit interaction) represents the interaction of the spins of electrons with the magnetic field generated by the orbital movement of the electrons. The second term H_{SR} (spin–rotation interaction) characterizes the interaction of the electrons' spins with the magnetic field generated by the orbital movement of the nuclei. The third term H_{SS} (spin–spin interaction) represents the interaction of the electrons' spins with each other. These terms are responsible for the fine structure observed in the electronic multiplet states of radicals or ions which have a lone electron or the excited or multiplet electronic states of the dioxygen fundamental state, O_2 [BRO 03].

Equation [1.6e] (H_{hfs}, *hfs*: hyperfine structure) describes the interaction of the nuclei's magnetic or quadrupolar moments with the intramolecular field. The first term H_{IL} (spin–orbit interaction) results from the interaction of nuclear spins with the magnetic field generated by the electrons' orbital movement. The second term H_{IR} (spin–rotation interaction) represents the interaction of the nuclear spins with the magnetic field generated by the nuclei's orbital movement. The third term H_{II} (spin–spin interaction) represents the interaction of nuclear spins with each other. The fourth term H_{IS} (spin–spin interaction) characterizes the interaction of the nuclear spins with the electrons' spins and the Fermi contact interaction. The fifth term H_{EQ} results from the interaction between the electric quadrupolar moments of the nuclei and the gradient of the intramolecular electric field. These interactions can be observed in high-resolution spectroscopy in the microwave frequency region.

If one takes the non-sphericity of the distribution of nuclear charges and their movements into account, other electrical and magnetic moments are also present that are represented by terms of the higher order in the expansion of the Hamiltonian.

As the Hamiltonian matrix is infinite, the exact resolution of the Schrödinger eigenvalues equation (equation [1.1]) is generally impossible. To determine the energy levels of a diatomic molecule, it is necessary to simplify the Hamiltonian.

By a judicious choice of the axes system, the Hamiltonian can be expressed in a form that allows the variables to be separated:

$$H_0 = H - H' = T_{CM} + H_{eVR} + H_{SS} + H_{II}, \qquad [1.7]$$

where H' represents the coupling terms and $H_{eVR} = H_e + H_V + H_R$ is the sum of Hamiltonians relative to the electrons and the nuclei's vibration and rotation movements.

The wave functions and the eigenenergies can then be expressed as follows:

$$|\Psi_0\rangle = |\Psi_{CM}\rangle \times |\Psi_e\rangle \times |\Psi_V\rangle \times |Jm_J\rangle \times |Sm_s\rangle \times |Im_I\rangle \quad [1.8]$$

$$E_0 = E_{CM} + E_e + E_V + E_R + E_{SS} + E_{II}, \quad [1.9]$$

where $\Psi_{CM} = \langle R_{CM} | \Psi_{CM} \rangle$ is the wave function associated with the translation movement of the CM represented by a plane wave, $\Psi_e = \langle ..., r_e, ... | \Psi_e \rangle$ is the wave function associated with the movement of the electrons that depends on the electrons' position, $\Psi_V = \langle q | \Psi_V \rangle$ is the wave function associated with the vibration movement of the nuclei around their equilibrium position and that depends on the normal coordinate q (section 2.2.2 of Chapter 2), $\Psi_R = \langle \theta, \phi | Jm_J \rangle$ is the wave function associated with the rotation movement of the overall molecule around its CM (section 2.2.2 of Chapter 2) and $|Sm_s\rangle$ and $|Im_I\rangle$ the wave functions of the electronic and nuclear spins, respectively.

In a first approximation, the molecule's energy is a sum of terms relating to the different degrees of freedom, which are translational (for the CM), electronic, vibrational and rotational (for classical ones) or electronic and nuclear spins (for quantum ones). No approximation is necessary for separating the movement of the CM (first Eckart conditions: the origin of the frame tied to the molecule is placed at the CM) and the spins' degrees of freedom. However, the decomposition of $H_{eVR} = H_e + H_V + H_R$ requires the application of BO approximation or the adiabatic approximation, which allows the nuclei's movements to be separated from those of the electrons. This approximation relies on the ratio of the masses of an electron and a proton (m_e/m_N~1/1,836), which justifies that the nuclei's movement is slower than those of the electrons under the effect of forces of the same order of magnitude. The electronic movement may then be processed for a fixed position of the nuclei.

The exact expression of the rovibronic wave function is given by:

$$|\Psi_{eVR}\rangle = |\Psi_e\rangle \left(|\Psi_V\rangle |\Psi_R\rangle + |\Psi_{VR}\rangle \right) + |\Psi_{eN}\rangle \quad [1.10]$$

where $|\Psi_{VR}\rangle$ is associated with the interaction between the vibration and the rotation and $|\Psi_{eN}\rangle$ with the interaction between the movement of the electrons and that of the nuclei. When these interactions are neglected, the wave function can be expressed as a product of the individual wave functions (equation [1.8]), the electronic wave function then depending parametrically on the internuclear distance.

The molecular energy variation corresponding to a transition between two energy levels is written for the rovibronic part as follows:

$$\Delta E_{eVR} = \Delta E_e + \Delta E_V + \Delta E_R \quad [1.11]$$

with $\Delta E_e >> \Delta E_V >> \Delta E_R$ (Figure 1.2), which corresponds to the electronic levels that are relatively spaced in comparison to the vibrational sublevels; the latter are themselves more spaced than the rotational sublevels.

These scales of magnitude enable the spectra to be classified into three categories:

– the rotation spectra in which the electronic energy and the vibration energy are unchanged ($\Delta E_e = 0$ and $\Delta E_V = 0$). The spectrum is shaped in lines, which are each associated with a value of ΔE_R;

– the vibration–rotation spectra in which the electronic energy is unchanged ($\Delta E_e = 0$). The spectrum is shaped in bands which are each associated with a value of ΔE_V. Each band is composed of lines associated with the different values of ΔE_R;

– the electronic spectra comprised of different systems of bands that are each associated with a value of ΔE_e. The bands of the same system correspond to different values of ΔE_V. Each band is composed of lines associated with the values of ΔE_R.

The rotation spectra are observed in the microwave and far-infrared ranges (between 25 and 1,000 µm or 400 and 10 cm^{-1}); the vibration–rotation spectra in the mid-infrared range (between 2.5 and 25 µm or 4,000 and 400 cm^{-1}) or near-infrared range (between 0.75 and 2.5 µm or 13,333 and 4,000 cm^{-1}); the electronic spectra in the near-infrared, visible or ultraviolet range. The time scales corresponding to these degrees of freedom are located between 0.1 and 1 fs for the electrons, between 10 and 100 fs for the vibration and between 0.1 and 10 ps for the rotation.

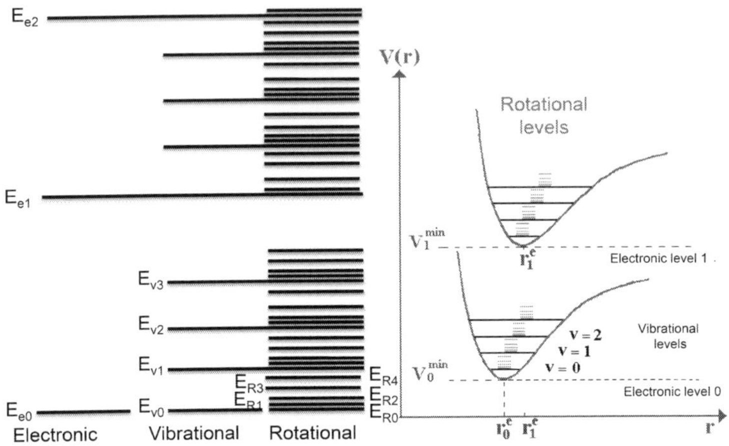

Figure 1.2. *Diagram of the electronic, vibrational and rotational energies of a molecule*

1.3. Symmetry properties of a diatomic molecule

1.3.1. *Group of symmetry*

A diatomic molecule may form in the gaseous phase when the electrons that are initially localized on each of the atoms are shared, resulting in a molecule's bonding state for an approximate distance corresponding to the molecule's bond length.

If one looks at a diatomic molecule's rovibronic degrees of freedom, comprising n electrons, the most general movement depends on $6 + 3n$ position coordinates. If we discard the three translation degrees of freedom (T_{CM}) of the CM which are not involved in the study of rovibronic spectra, this number is reduced to $3 + 3n$ coordinates, namely:

– $3n$ electronic coordinates;

– $6 - 5 = 1$ vibrational coordinates (5:3 for the translation and 2 for the rotation);

– two rotational coordinates.

We represent a molecule by defining a geometric figure corresponding to the equilibrium configuration of the two nuclei. The reference system connected to the equilibrium configuration (section 2.2 of Chapter 2) is chosen so that its origin is at the CM (first Eckart condition) and its axes along the principal directions of inertia of the equilibrium configuration. In the case of a diatomic molecule and a general linear molecule, the axis z is aligned on the internuclear axis and the two other axes are chosen to be perpendicular to z and to each other and have an arbitrary orientation. At a given instant, the configuration that corresponds to the physical situation differs from the reference configuration, which is arbitrarily set. The movement of the two nuclei is decoupled into two movements: a block rotation of the reference configuration (xyz mobile frame moving with the molecule) with respect to a fixed frame (laboratory-fixed referential, $(O,\vec{X},\vec{Y},\vec{Z})$) defined by two Euler angles θ and φ (section 2.3.2 of Chapter 2) and the vibration defined by the instantaneous positions of the two nuclei with respect to their equilibrium position. The axis of the reference configuration coincides with the axis of the instantaneous configuration and the vibration is described by a normal coordinate q (section 2.2.2 of Chapter 2) equal to the product of the bond length by the square root of the reduced mass.

A diatomic molecule's symmetry properties are determined with reference to the symmetry point group of the nuclei's equilibrium configuration [AMA 80a, CAM 88, LAN 75, WIL 80]. First, we define a reference system connected to the equilibrium configuration and which moves with it. We then establish the symmetry point group by identifying the different possible reference system's changes for which the equilibrium configuration is superimposed. The operations that bring about such changes are the symmetry operations.

The configuration of the homonuclear diatomic molecules A_2, with identical atoms, comprises indiscernible nuclei (A, A). Its corresponding point group $D_{\infty h}$ (Table 1.1) comprises the following symmetry operations (Figure 1.3):

– an infinity of rotations $C_z(\varphi)$ of an angle φ around the axis z including the identity operation I;

– an infinity of improper rotations $S_z(\varphi)$ (rotation of an angle φ around the axis z and inversion with respect to the bond center), including the reflection operation with respect to the plane xy (rotation of an angle 0 around the axis z and inversion with respect to the bond center);

– an infinity of rotations $C_t(\pi)$ of an angle π around the axes t perpendicular to z and passing through the origin;

– an infinity of reflections σ_{zt} with respect to the planes zt containing the axis z and making any angle with the plane zx.

The symmetry elements of A_2 are, respectively, the rotation axis of the first kind z (proper axis that includes the identity), the rotation axes t, the symmetry planes zt and the rotation axis of the second kind z (improper axis that includes the inversion).

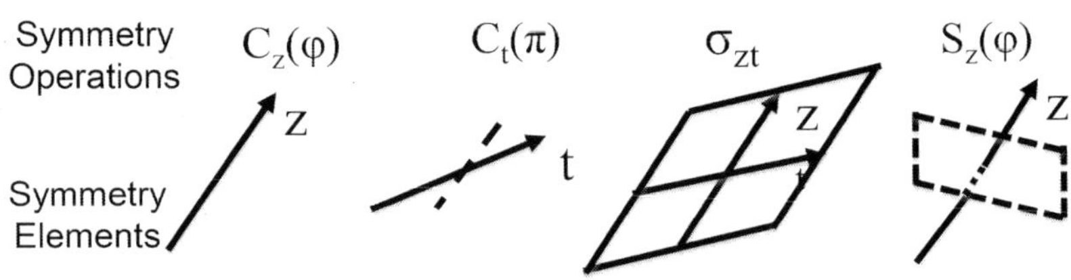

Figure 1.3. *Symmetry operations (Schoenflies notation) and symmetry elements of a homonuclear diatomic molecule*

	I	$2C_\infty(\varphi)$	$^\infty\sigma_V$	I	$2S_\infty(\varphi)$	$^\infty C_2$	
Σ_g^+	1	1	1	1	1	1	q, Ψ
Σ_u^+	1	1	1	–1	–1	–1	μ_Z, T_z
Σ_g^-	1	1	–1	1	1	–1	J_Z
Σ_u^-	1	1	–1	–1	–1	1	
Π_g	2	2cosφ	0	2	–2cosφ	0	(μ_X, μ_Y) (T_X, T_Y)
Π_u	2	2cosφ	0	–2	2cosφ	0	(J_X, J_Y)
Δ_g	2	2cosφ	0	2	2cosφ	0	
Δ_u	2	2cosφ	0	–2	–2cosφ	0	

Table 1.1. *Symmetry Group $D_{\infty h}$: table of characters*

For the heteronuclear molecules AB, with distinct atoms, the symmetry point group $C_{\infty v}$ (Table 1.2) comprises the following symmetry operations:

– an infinity of rotations $C_z(\varphi)$ of an angle φ around the axis z including the identity operation I;

– an infinity of reflections σ_{zt} with respect to the planes zt containing the axis z and making any angle with the plane zx.

The symmetry elements of AB are, respectively, the rotation axis of the first species z and the symmetry planes zt.

	I	$2C_\infty(\varphi)$	$\infty\sigma_V$	
Σ^+	1	1	1	μ_Z, T_Z, q, Ψ
Σ^-	1	1	−1	J_Z
Π	2	$2\cos\varphi$	0	(μ_X, μ_Y) (T_X, T_Y) (J_X, J_Y)
Δ	2	$2\cos\varphi$	0	
Φ	2	$2\cos\varphi$	0	

Table 1.2. *Symmetry group $C_{\infty V}$: table of characters*

The definition of the symmetry types of the groups $C_{\infty v}$ and $D_{\infty h}$ is connected to the behavior of the physical quantities and the other molecular properties with respect to the symmetry operations. The symbols Σ, Π, Δ... are related to the behavior with respect to the rotation $C_z(\varphi)$; there are two one-dimensional irreducible representations with totally symmetric cylindrical symmetry (Σ) and the others, which are all two-dimensional, are in an infinite number (Π,Δ,Φ,...); the sign + or – added as a superscript to a symbol Σ (electronic distribution possessing the cylindrical symmetry around z or invariant states in a rotational symmetry operation around z) is related to the symmetric or antisymmetric character with respect to a reflection σ_{zt}. For the group, $D_{\infty h}$, the index g (gerade or even) or u (ungerade or uneven) is related to the symmetric or antisymmetric character with respect to the inversion. The symmetry operations of the group $D_{\infty h}$ are twice those of the group $C_{\infty V}$ as a result of the existence of the inversion operation.

As a result, the irreducible representations of the symmetry point group of the diatomic molecule, homonuclear molecule ($D_{\infty h}$) or heteronuclear molecule ($C_{\infty v}$), are, respectively, designated by $\Sigma_g^+, \Sigma_u^+, \Sigma_g^-, \Sigma_u^-$, $\Pi_g, \Pi_u, \Delta_g, \Delta_u, \Phi_g, \Phi_u, \Gamma_g, \Gamma_u$.... and $\Sigma^+, \Sigma^-, \Pi, \Delta, \Phi, \Gamma$.... and are used to highlight and thus identify the symmetry properties of the rovibronic states.

The symmetry of the states bear particular significance on the possible transitions that may take place between the quantum states and which are responsible for the spectra observed. A transition between two quantum states takes place if at least one of the integrals of the type:

$$R_{mn} = \left\langle \Psi_{eVR}^m \middle| R \middle| \Psi_{eVR}^n \right\rangle = \int \Psi_e^{m*} \Psi_V^{m*} \Psi_R^{m*} R \Psi_e^n \Psi_V^n \Psi_R^n d\tau \qquad [1.12]$$

is different from zero and is therefore symmetric with respect to the symmetry operations, which leave the molecule invariant. This is equivalent to saying that $\Gamma(|\Psi_m\rangle) \otimes \Gamma(|\Psi_n\rangle) \otimes \Gamma(R)$ must contain the unity representation, that is the symmetry of the product of the two states between which the transition occurs must be of the same type of symmetry as that of the operator R.

In infrared spectroscopy, corresponding to the absorption or emission of photons (absorption or emission spectroscopy), the operator R is one of the components (μ_x, μ_y, μ_z) of the electric dipole moment (sections 2.2.3, 2.3.3, and 2.4.2 of Chapter 2). In elastic diffusion (Rayleigh, Mie) or inelastic diffusion (Raman, Brillouin), R is one of the operators connected to the elements $\alpha_{xx}, \alpha_{yy}, \alpha_{zz}, \alpha_{xy}, \alpha_{yz}, \alpha_{xz}$ of the molecule's polarizability tensor. The component of the permanent electric dipole moment μ_z of a heteronuclear diatomic molecule is of the type Σ^+. The electric dipole moment induced by the vibration (section 2.2.3 of Chapter 2) has the same symmetry as that of the power of the vibration normal coordinate, which is also of the type Σ^+.

The homonuclear diatomic molecules have no permanent or induced electric dipole moment and therefore do not have a vibration–rotation spectrum. This situation implies that the molecular detection based on rotational transitions does not apply to hydrogen, for example. However, there is so much hydrogen present in the universe that, although the quadrupolar transitions are weak, the corresponding spectrum is detectable, like that observed in the region of Orion where stars are formed.

1.3.2. *Symmetry of the electronic states*

To label the electronic states of a diatomic molecule, one uses the same convention as for the atoms, with Latin or Greek letters in upper case $(\vec{L},\vec{S},\Sigma,\Pi,\Delta..)$ to characterize all the electrons and in lower case $(\vec{l},\vec{s},\sigma,\pi,\delta...)$ to characterize an electron individually [AMA 80a, AMA 80b, HER 50, HOL 86, LAN 75].

In an atom, the orbital angular momentum $\vec{L}$ of all the electrons keeps a fixed direction in space due to the spherical symmetry of the nucleus' force field (Figure 1.4(a)). In a diatomic molecule, the force field possesses a cylindrical symmetry around the internuclear axis z and consequently, the orbital angular momentum $\vec{L}$ of all the electrons precesses around z just like an atom placed in an electrical field (Figure 1.4(b)). The projection of $\vec{L}$ on z is equal to $\hbar M_L$ with $M_L = -L,-(L-1),.......,(L-1),L$. L is a quantum number that is connected to the modulus of $|\vec{L}|$ equal to $\hbar\sqrt{L(L+1)}$. Two states with values $\hbar M_L$ and $\hbar M_L^{'}$ have the same energy, which thus depends on the value $\Lambda = |M_L|$. The orbital angular momentum of the electrons is therefore characterized by the axial vector $\vec{\Lambda}$ directed along the z axis, that is the component of $\vec{L}$ along z, rather than $\vec{L}$ itself. Depending on the values of $\Lambda = 0,1,2,3...$, the electronic state is designated by $\Sigma,\Pi,\Delta,\Phi....$ A state Σ is not degenerate, whereas the states $\Pi,\Delta,\Phi...$ are doubly degenerate.

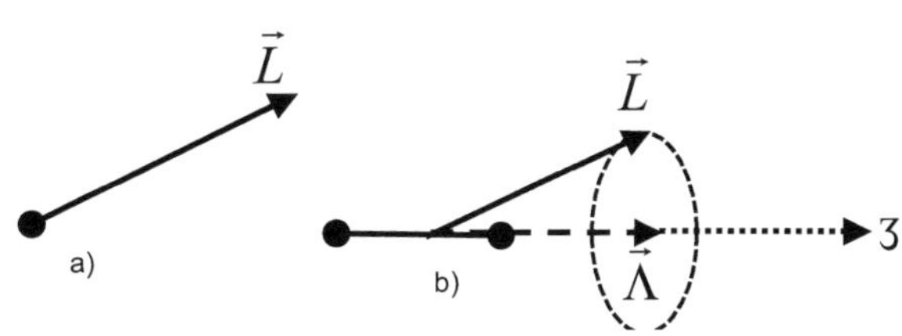

Figure 1.4. *Angular momentum of the electrons: a) atomic and b) diatomic*

The total spin of all the electrons noted $\vec{S}$ is not perturbed by an electrical field but only by a magnetic field. In a state Σ , the orbital movement has no magnetic moment so that the magnetic field is zero, which means that $\vec{S}$ remains fixed in space. In a state $\Pi, \Delta, \Phi...$, the orbital movement of the electrons gives rise to a magnetic field parallel to z. A precession of $\vec{S}$ around z (Figure 1.5) then follows. The projection of $\vec{S}$ on z is equal to $\hbar S_z$ with $S_z = -S, -(S-1), ..., (S-1), S$. S is the quantum number that is connected to the modulus of $|\vec{S}|$ equal to $\hbar\sqrt{S(S+1)}$. For each value of S, there are $(2S+1)$ values of S_z. The precession is due to a magnetic coupling and states corresponding to the opposite values of S_z possess different energies. The quantum numbers S and S_z take integral or half-integral values depending on whether the molecule contains an even number or an odd number of electrons (to avoid confusion, we use S_z instead of Σ as is customary).

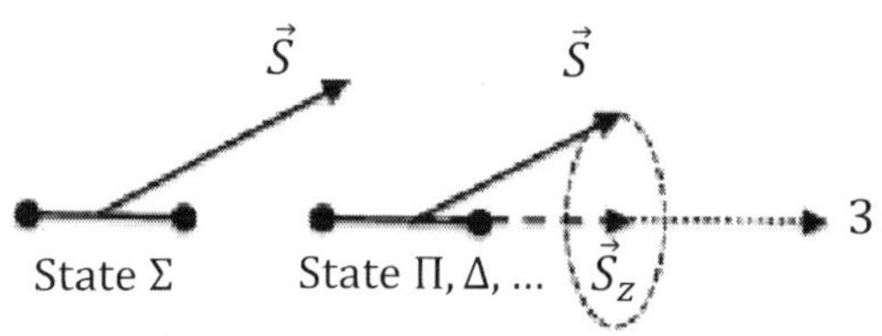

Figure 1.5. *Spin of electrons for different electronic states (Σ, Π, Δ ...)*

The total angular momentum of the electrons is obtained by the addition of $\vec{L}$ and $\vec{S}$. Its projection on the z axis is equal to $\hbar\Omega$, such that $\Omega = |\Lambda + S_z|$ (Figure 1.6). Ω takes integral or half-integral value just like S_z depending on whether the molecule contains an even or odd number of electrons. An electronic state is represented by the symbol $^{2S+1}[\Gamma]_{\Lambda+S_z}$, where $[\Gamma]$ designates the symbol $\Sigma, \Pi, \Delta, \Phi, ...$ associated with the value 0, 1, 2 or 3 of Λ, respectively. The definition of a state is completed by the multiplicity 2S+1 at the left superscript and at the right subscript by the projections of the total angular momentum of the electrons. For each value of Λ and S, there are $2S+1$ states corresponding to the different values of S_z, which define the orientations of $\vec{S}$ with respect to $\vec{\Lambda}$ and which correspond to the multiplicity of the electronic term.

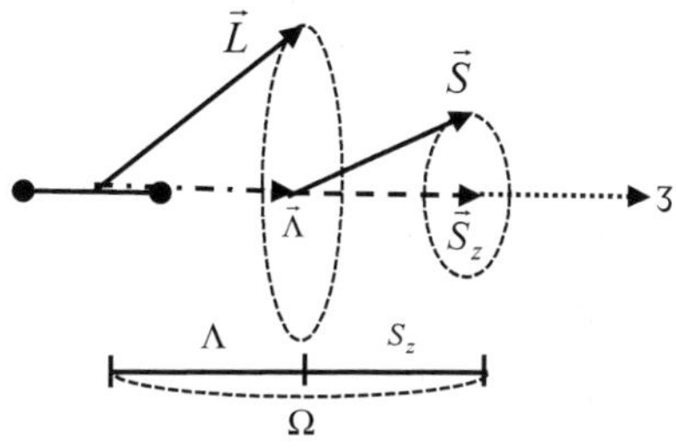

Figure 1.6. *Angular orbital momentum and spin of electrons and total angular momentum of the electrons*

For example, for $\Lambda = 0$ and $S = 1$, we have $S_z = -1, 0, +1$ (multiplicity of 3), and $\Lambda + S_z = -1, 0, +1$. The electronic term is a triplet $^3\Sigma$ whose components are noted as $^3\Sigma_{-1}, ^3\Sigma_0, ^3\Sigma_1$. For $\Lambda = 1$ and $S = 1/2$, we have $S_z = -1/2, +1/2$ (multiplicity of 2) and $\Lambda + S_z = +1/2, +3/2$. The electronic term is a doublet $^2\Pi$ whose components are noted as $^2\Pi_{1/2}, ^2\Pi_{3/2}$ (Figure 1.7).

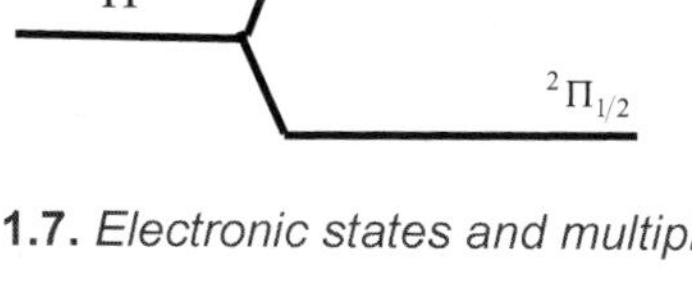

Figure 1.7. *Electronic states and multiplicities*

The symbols $\Sigma, \Pi, \Delta, \Phi, \ldots$ can be defined from the molecule's symmetry properties without reference to the angular momentum value Λ. The states

Ψ, which are invariant in a rotational symmetry operation of an angle φ around the axis z, are of the type Σ. The other states are doubly degenerate and are studied in pairs $\{\Psi_{-1}, \Psi_{+1}\}$. The trace Tr of the square matrix 2 × 2, of a rotational symmetry operation of angle φ around the axis z of the basis $\{\Psi_{-1}, \Psi_{+1}\}$ allows the symmetry type of the doubly degenerate states to be defined. This trace is expressed as $Tr = 2\cos m\varphi$ and following the value of $m = 1, 2, 3, \ldots$, the states $\{\Psi_{-1}, \Psi_{+1}\}$ are of the type $\Pi, \Delta, \Phi, \ldots$. Depending on whether the function is symmetric or antisymmetric with respect to the reflection symmetry operations of the plane containing the internuclear axis and in the case of a homonuclear diatomic molecule, with respect to the center of symmetry, the states are distinguished by the use of symbols +, – and g, u, respectivcly.

1.3.3. *Symmetry of the total wave functions*

The molecular states are described by total wave functions that are, in a first approximation, the product of translational, electronic, vibrational, rotational and spin wave functions (equation [1.8]: $|\Psi_0\rangle = |\Psi_{CM}\rangle \times |\Psi_e\rangle \times |\Psi_V\rangle \times |Jm_J\rangle \times |Sm_s\rangle \times |Im_I\rangle$). Because of the first Eckart condition, with the origin of the mobile frame at the nuclear CM, the overall translation of the molecule is not considered, which allows the translational degrees of freedom to be rigorously separated from the other ones.

The electronic wave function $|\Psi_e\rangle$ depends on the electrons' position coordinates ($\langle \vec{r}_e | \Psi_e \rangle$), the vibration wave function $|\Psi_V\rangle$ depends on the vibration's normal coordinate ($\langle q | \Psi_V \rangle$), a linear combination of the weighted Cartesian displacements of the nuclei, the rotation wave function $|\Psi_R\rangle \equiv |J, M\rangle$ depends on Euler angles ($\langle \theta, \varphi | J, M \rangle$) and the nuclear spin's wave function, which is of quantum nature, depends on the nucleus' mass number M (even M: integral spin obeying the Bose–Einstein statistics; odd M: half-integral spin obeying the Fermi–Dirac statistics) (Table 1.6).

To determine the symmetry of an electronic or vibrational wave function, one must know how the electrons' coordinates and the vibrational normal coordinate transform in a symmetry operation [AMA 80a, CAM 88,

LAN 75, WIL 80]. In that case, the symmetry operation is defined as a change in the xyz reference frame accompanied by a permutation of the indices that label identical nuclei. The reference system is rotated with respect to the nuclei's equilibrium or instantaneous configuration.

In a diatomic molecule, the internal electrons are localized around each of the molecule's nuclei on complete shells: the orbital angular momentum and the spin momentum are then equal to zero. The valence electrons that contribute to the chemical bond are generally grouped into pairs with parallel spins ($S = 0$) on the orbitals, which have an orbital angular momentum equal to zero ($\Lambda = 0$). The fundamental state is a singlet ($^1\Sigma$), although there are exceptions to this rule, such as for nitrogen monoxide NO, a heteronuclear diatomic molecule whose fundamental electronic level is a doublet $^2\Pi$ ($\Lambda = 1$ and $S = 1/2$), and dioxygen O_2, a homonuclear diatomic molecule, whose fundamental electronic state is a triplet $^3\Sigma$ ($\Lambda = 0$ and $S = 1$). In this case, it is shown that the fundamental electronic state of a stable molecule is totally symmetric with respect to the symmetry operations of a diatomic molecule, and thus either of the type Σ (heteronuclear) or of the type Σ_g^+ (homonuclear). For NO, the state is of the type Π and for O_2, it is of the type Σ_g^- (Table 1.3). The lowest vibrational level is always totally symmetric because the vibrational wave function depends on the internuclear distance, which is invariant with respect to all the group's symmetry operations, that is of the type Σ (heteronuclear) or of the type Σ_g^+ (homonuclear) (Table 1.3).

Symmetry of the fundamental level	Homonuclear group $D_{\infty h}$	Heteronuclear group $C_{\infty V}$
Electronic	Σ_g^+ Except O_2 (Σ_g^-)	Σ^+ Except NO (Π)
Vibrational	Σ_g^+	Σ^+

Table 1.3. *Symmetry type of the electronic or vibrational fundamental state*

In the case of rotation, the equilibrium or instantaneous configuration of the nuclei (the object) is rotated with respect to the XYZ fixed reference system. For molecules belonging to the symmetry group $D_{\infty h}$, the exchange effect should also be considered with respect to the identical nuclei and is determined by rotating the reference frame tied to the equilibrium configuration around the y axis. This operation must be performed in the three subspaces, electronic, vibrational and rotational. In the electronic and vibrational subspaces, the reference system is changed (symmetry operation $C_y(\pi)$) without modifying the object studied, that is the instantaneous configuration of the electrons and the nuclei.

Symmetry operation	Level	Character
$C_y(\pi) \equiv \sigma_{zx} \bullet i$	Vibrational Σ_g^+	Ψ_V symmetric
$C_y(\pi)$	Electronic Σ_g^+ Electronic Σ_u^-	Ψ_e symmetric
$C_y(\pi)$	Electronic Σ_u^+ Electronic Σ_g^-	Ψ_e antisymmetric
$C_y(\pi)$	Even rotational J Odd rotational J	Ψ_R symmetric Ψ_R antisymmetric

Table 1.4. *Symmetry type of the electronic, vibrational and rotational level with the exchange of the two identical nuclei*

In the case of rotational subspaces, the studied object (equilibrium configuration of the nuclei) is rotated without modifying the XYZ fixed reference system. Consequently, the xyz system plays the role of the reference system in the electronic and vibrational subspaces, and in the rotational subspace, that of the studied object (Table 1.4). We use the symbols "s" and "a" to characterize the rotational levels symmetric and antisymmetric with respect to the exchange of identical nuclei.

Another type of characterization is applied to all diatomic (or linear) molecules which either belong to the $C_{\infty v}$ symmetry group or to the symmetry $D_{\infty h}$ group. The total wave function $|\Psi_t\rangle$ is labeled with the index "+" or "-" depending on whether it remains invariant or changes sign when the xyz reference frame (right-handed) is replaced by a frame in the opposite direction (left-handed). In this case, we perform the symmetry operation σ_{zx} in the electronic or vibrational subspaces. In the rotational subspace, this operation is not possible. In this case, the fixed XYZ frame and the mobile frame xyz must be simultaneously inverted, which is impossible. Then the symmetry operation $C_y(\pi)$ must be applied to the mobile frame xyz (the fixed XYZ frame having been inverted). It can be verified that the operation $C_y(\pi)$ transforms the Euler angles (section 2.3 of Chapter 2) such as: $\theta \rightarrow \pi - \theta$ and $\phi \rightarrow \phi + \pi$. The rotational wave function $|\Psi_R\rangle$ transforms in the following way $(-1)^J|\Psi_R\rangle$, meaning that it is invariant if J is even and changes sign for odd J (Table 1.5).

Symmetry operation	Homonuclear Group $D_{\infty h}$	Heteronuclear Group $C_{\infty V}$	Character
σ_{zx}	Vibrational Σ_g^+	Vibrational Σ^+	Ψ_V symmetric
σ_{zx}	Electronic Σ_g^+ Electronic Σ_u^+	Electronic Σ^+ Electronic Σ^+	Ψ_e symmetric
σ_{zx}	Electronic Σ_g^- Electronic Σ_u^-	Electronic Σ^- Electronic Σ^-	Ψ_e antisymmetric
$C_y(\pi)$	Even rotational J Odd rotational J	Ψ_R symmetric Ψ_R antisymmetric	Ψ_R symmetric Ψ_R antisymmetric

Table 1.5. *Symmetry type of the electronic, vibrational and rotational level in an inversion operation of the xyz frame*

These symmetry properties are useful for obtaining the selection rules in emission or absorption spectroscopy, on the one hand, and in Raman diffusion spectroscopy, on the other hand, and thus to determine the relevant technique to use in order to detect a diatomic molecule. In absorption or emission spectroscopy, it is shown that the transitions are only possible between states of different symmetry $(+ \leftrightarrow -)$ and that in Raman diffusion spectroscopy the transitions are only possible between states of the same symmetry $(+ \leftrightarrow + \; or \; - \leftrightarrow -)$.Thus, for homonuclear molecules, only the Raman diffusion is possible with transitions between states of the same symmetry $(s \leftrightarrow s \; or \; a \leftrightarrow a)$ corresponding to a variation of the quantum number J, which must be even ($\Delta J = 2$). Symmetry properties show that it is impossible to observe the pure rotation spectrum in emission or absorption for a homonuclear diatomic molecule.

To analyze the sprectra observed in emission (respectively in absorption), and in Raman diffusion, it is necessary to take into account the spins of the electrons and of the nuclei by considering the total wave function (equation

[1.8]) to establish selection rules. Generally, this wave function is written by including the spin variables of electrons in the electronic wave function, such that only the exchange effect of the identical nuclei on the symmetry of the total wave function needs to be considered. The wave function is then written under the form: $|\Psi_t\rangle = (|\Psi_e\rangle \times |Sm_s\rangle \times |\Psi_V\rangle \times |Jm_J\rangle) \times |\mathrm{Im}_I\rangle$. It is obvious that only the spectra of homonuclear molecules will show structures that are directly connected to the possibility of exchanging the identical nuclei in the molecule. This effect may be determined by using the permutation operator properties of the two particles. In a permutation operation, the total wave function may be invariant or may change its sign, which corresponds to the two eigenvalues –1 and +1 of the permutation operator of the two particles. The wave functions can be grouped into "*s*" states and "*a*" states depending on whether the wave function is symmetric (+1) or antisymmetric (–1) in a permutation operation of indiscernible particles. The effect of the operator O associated with the physical quantity that induces the transition between two states, resulting in the spectra that are observable in emission or absorption spectroscopy (electric dipole moment, electric quadrupole moment, magnetic dipole moment), in Raman diffusion spectroscopy (polarizability) or in collision spectroscopy (collision operator) can be calculated from the transition moment $\langle \Psi_t' | O | \Psi_t'' \rangle$. It is shown that because the operator O is invariant in an exchange operation of two identical particles, the product $\langle \Psi_t' | \Psi_t'' \rangle$ must also be invariant which implies that $|\Psi_t'\rangle$ and $|\Psi_t''\rangle$ are either both symmetric or both antisymmetric. Consequently, a system that is initially in a symmetric state or in an antisymmetric state remains as such. The states may thus be grouped into two categories.

In the case of the exchange of two electrons in one atom, a state $|\Psi_t\rangle$ is always antisymmetric according to the Pauli principle, which stipulates that two electrons cannot be found in states characterized by identical values of all the quantum numbers. Two electrons on the same orbital (same space variables) must have different spin orientations ($\pm 1/2$). In the case of a nucleus of mass number M which contains M particles of spin $1/2$, exchanging two nuclei amounts to exchanging M times particles of spin $1/2$. The wave function $|\Psi_t\rangle$ changes sign each time the two particles are exchanged, so that $|\Psi_t\rangle$ is symmetric or antisymmetric depending on whether M is even or odd. The nuclear spin wave function which is of quantum origin depends on the nucleus' mass number M (even M: integral spin obeying the Bose–Einstein statistics; odd M: half-integral spin obeying the Fermi–Dirac statistics) (Table 1.6). These different properties are responsible for the structures present in the IR and Raman spectra of the homonuclear diatomic molecules.

In the case of a homonuclear diatomic molecule A-A, the nuclear spin wave function can be put under the form of a product: $|\Psi_I\rangle = |\Psi_i^{(1)}\rangle \times |\Psi_j^{(2)}\rangle$, an expression which is interpreted with the nucleus numbered 1 in the spin state *i* and the nucleus numbered 2 in the spin state *j*. The states *i* and *j* are characterized by the value of the quantum number I_z which may take $2I+1$ values ($-I$, $-(I-1)$, etc., $I-1$, I), where *I* designates the nucleus' spin.

For example for $I=0$, we have $I_z=0$, and the wave function is written as: $|\Psi_I\rangle = |\Psi_0^{(1)}\rangle \times |\Psi_0^{(2)}\rangle$. This wave function is symmetric in the exchange of nuclei.

For $I=1/2$, we have $I_z = -1/2, +1/2$. Therefore, four wave functions can be written (multiplicity of 4): $|\Psi_1\rangle = |\Psi_{1/2}^{(1)}\rangle \times |\Psi_{1/2}^{(2)}\rangle$, $|\Psi_2\rangle = |\Psi_{1/2}^{(1)}\rangle \times |\Psi_{-1/2}^{(2)}\rangle$, $|\Psi_3\rangle = |\Psi_{-1/2}^{(1)}\rangle \times |\Psi_{1/2}^{(2)}\rangle$ and $|\Psi_4\rangle = |\Psi_{-1/2}^{(1)}\rangle \times |\Psi_{-1/2}^{(2)}\rangle$. The wave functions $|\Psi_1\rangle$ and $|\Psi_4\rangle$ are symmetric. From the two other wave functions, $|\Psi_2\rangle$ and $|\Psi_3\rangle$, two other wave functions can be formed, one of which is symmetric, $|\Psi_s\rangle = 1/\sqrt{2}\left(|\Psi_2\rangle + |\Psi_3\rangle\right)$ and the other which is antisymmetric $|\Psi_a\rangle = 1/\sqrt{2}\left(|\Psi_2\rangle - |\Psi_3\rangle\right)$. Therefore, when $I = 1/2$, four wave functions are obtained, three symmetric and one antisymmetric.

Statistic	Number of mass M	Nuclear spin	Character
Bose–Einstein	Even	Integral	C_t symmetric
Fermi–Dirac	Odd	Half-integral	C_t antisymmetric

Table 1.6. *Nuclei's statistics of the even and odd mass number and wave function symmetry*

In general, it can be shown that for a nucleus' spin *I*, there are $(2I+1)(I+1)$ symmetric functions and $(2I+1)(I)$ antisymmetric functions. The symmetric states are called *ortho* states and the antisymmetric states are called *para* states. There are always more *ortho* states than *para* states. When $I = 0$, there is no *para* state. A state can also be characterized from the addition of the spins of each nucleus that we label *T*, such that $|I_1 - I_2| \leq T \leq |I_1 + I_2|; (T = 0, 1, \ldots .2I)$ and the projection of *T* on the internuclear axis $-T \leq M_T \leq +T; \Delta M_T = +1$. It is shown that the *ortho* states correspond to the even *T* states and the *para* states correspond to the odd *T* states, which are valid, regardless of the statistics: Bose–Einstein or Fermi–Dirac.

When one considers the total wave function $|\Psi_t\rangle = |\Psi_{eVR}\rangle |\Psi_I\rangle$, the product must be symmetric if the nuclei obey the Bose–Einstein statistics and must be antisymmetric if the nuclei obey the Fermi–Dirac statistics. As a result, the relative statistical weight of the $|\Psi_{eVR}\rangle_s$ states of the "*s*" type and the $|\Psi_{eVR}\rangle_a$ states of the "*a*" type depend on this property. We have $\frac{|\Psi_{eVR}\rangle_s}{|\Psi_{eVR}\rangle_a} = \frac{I+1}{I}$ if the nuclei obey the Bose–Einstein statistics and $\frac{|\Psi_{eVR}\rangle_s}{|\Psi_{eVR}\rangle_a} = \frac{I}{I+1}$ if the nuclei obey the Fermi–Dirac statistics. If $I = 0$, only the states $|\Psi_{eVR}\rangle_s$ are present.

These results are summarized in Table 1.7.

Statistics	Ψ_{eVR}	Ψ_I	Statistical weight
Bose–Einstein (*I* integer) Ψ_t symmetric	*s* *a*	*ortho* *para*	(2*I*+1)(*I*+1) (2*I*+1)*I*
Fermi–Dirac (*I* half-integer) Ψ_t antisymmetric	*s* *a*	*para* *ortho*	(2*I*+1)*I* (2*I*+1)(*I*+1)

Table 1.7. *Statistic weight of the "s" and "a" states for the nuclei with even and uneven mass number*

As the character "*s*" or "*a*" of the states depends on the parity of the rotational quantum number *J*, there is a correspondence between (*s, a*) and (even *J*, odd *J*), which depends on the symmetry type of the fundamental electronic level (Table 1.5). The rotational sublevels have different statistical weights depending on whether the *J* is even or odd.

For example, for the diatomic molecules N_2, D_2, H_2, and O_2, the exchanged spins are, respectively, 1 for N and D, 0 for O and ½ for H. For N_2, which has a fundamental electronic state of the Σ_g^+ type, the even *J* levels are "*s*" and odd *J* are "*a*". As the spin of N is equal to 1, the "*s*" levels are *ortho* and the "*a*" levels are *para*. As the ratio of the statistical weight of *s/a* is equal to 2/1 (Table 1.7), the even *J* levels are twice as populated as the odd *J* levels. We have the same result for D_2. For H_2, the fundamental electronic state is of the Σ_g^+ type, but the nuclei obey the Fermi-Dirac statistics, and the "*s*" levels are *para* and the "*a*" levels are *ortho*. As the ratio of the statistical weight *s/a* is equal to 1/3 (Table 1.7), this time the odd *J* levels are three times as populated as the even *J* levels. For O_2, the fundamental electronic state is of the Σ_g^- type (antisymmetric: Table 1.5) and as the nuclear spin is zero, only the *ortho* states are present and consequently only the odd *J* levels are populated (Table 1.5).

These symmetry considerations show that in the rotational Raman spectrum (LIDAR spectroscopy) of homonuclear diatomic molecules, the series of lines called Stokes and the series of lines called anti-Stokes have intensity alternations; one line out of two is more intense than its neighbor. For instance, in the N_2 spectrum, the lines corresponding to even *J* are twice as strong as the lines corresponding to odd *J*. For H_2, the lines corresponding to odd *J* are three times stronger than the even *J* lines. For O_2, the lines with even *J* are missing in the spectrum.

1.4. Example of the diatomic molecule with two electrons H_2, HD, D_2

1.4.1. *Hamiltonian of the isotopologues*

The most abundant molecule in the universe is dihydrogen. This is the simplest neutral molecule, composed of two nuclei and two electrons. This molecule is homonuclear and therefore belongs to the symmetry group

($D_{\infty h}$). If one considers the isotopologues with the deuterium D nucleus, the HD species is from the symmetry group ($C_{\infty v}$), whereas D_2 is from the symmetry group ($D_{\infty h}$). Dihydrogen can be detected in regions where the flux of ultraviolet photons is sufficiently screened to limit the dissociation of the molecule. H_2 is the main molecule that contributes to the mass of galaxies' dense gases and plays a fundamental role in astrophysics, particularly in the processes that regulate star formation and the evolution of galaxies. The physical–chemistry of the interstellar gases is initiated by the formation of H_2 on dust grains and it is dihydrogen that mainly contributes to the cooling of the astrophysical media via collisional processes. In the presence of radiation in the far UV range, H_2 is electronically excited and relaxes radiatively or non-radiatively to the fundamental electronic level, thus populating the high vibrational energy levels, which then emit fluorescence in the optical and infrared range. The electronic transitions of H_2 in UV cannot be detected from instruments based on the ground and require platforms above the earth's atmosphere because the latter is opaque to UV radiation. As H_2 is symmetric, the transitions are due to the very weak quadrupolar moment. In the case of the HD isotopologue, a dipole moment is present because the barycenter of negative charges does not coincide with that of the nuclei and, although very weak, it allows this species to be detected in the infrared. Moreover, the difference of nuclear masses is responsible for a breakdown of the electronic symmetry (g-u parity) in an inversion of electronic coordinates with respect to the geometric center of the nuclei. Interactions between the electronic states that have different symmetries are then possible and transitions forbidden with the H_2 or D_2 isotpologues are then possible.

The formulation of the Hamiltonian of a diatomic molecule is different if the origin of the reference frame connected to the molecule is placed at the nuclei's CM or if it is placed at the geometric center of the molecule. If we focus on the electronic spectra, it is more judicious to place the origin of the molecular reference frame at the geometric center. However, if our interest is on the vibration–rotation spectra when the molecule is in its fundamental electronic state, it is wiser to place the origin of the molecular reference frame at the nuclei's CM.

In a reference frame whose origin is at the geometric center of the diatomic molecule, the Hamiltonian is expressed as:

$$H_{eTVR} = T_{CM} + H_e + H_N + H_{pm} + H_{gu} \qquad [1.13a]$$

with

$$T_{CM}=-\frac{\hbar^2}{2M}\nabla^2_{CM} \quad [1.13b]$$

$$H_e=\sum_{e=1}^{2}\left\{-\frac{\hbar^2}{2m_e}\nabla^2_e+V\left(\vec{r}_e,\vec{R}_N\right)\right\} \quad [1.13c]$$

$$H_N=-\frac{\hbar^2}{2m_N}\nabla^2_{R_N} \quad [1.13d]$$

$$H_{pm}=-\frac{\hbar^2}{8m_N}\sum_{e,e'=1}^{2}\vec{\nabla}_e\vec{\nabla}_{e'} \quad [1.13e]$$

$$H_{gu}=-\frac{\hbar^2}{2\mathrm{m}_\alpha}\vec{\nabla}_{R_N}\sum_{e=1}^{2}\vec{\nabla}_e\,, \quad [1.13f]$$

where $m_N=\frac{M_1M_2}{M_1+M_2}$ is the reduced mass of the nuclei with mass M_1 and M_2, and $\mathrm{m}_\alpha=\frac{M_1M_2}{M_1-M_2}$. With respect to the reference frame whose origin is the CM of the nuclei, an additional term H_{gu} is present and disappears when the nuclei have the same mass and the term H_{pm} connected to the mass polarization is divided by 4. In this frame, the term H_{gu} explicitly highlights the difference between the H_2 and HD isotopologues.

If we explicate the electrostatic interaction term, $\sum_{e=1}^{2}V\left(\vec{r}_e,\vec{R}_N\right)$, we can write that:

$$\sum_{e=1}^{2}V\left(\vec{r}_e,\vec{R}_N\right)=\frac{e^2}{4\pi\varepsilon_0\left|\vec{r}_1-\vec{r}_2\right|}+\frac{e^2}{4\pi\varepsilon_0\left|\vec{R}_1-\vec{R}_2\right|}-\sum_{i=1}^{2}\sum_{k=1}^{2}\frac{e^2}{4\pi\varepsilon_0\left|\vec{R}_i-\vec{r}_k\right|} \quad [1.14]$$

which allows the electronic and nuclear Hamiltonians to be rearranged under the following form, by disregarding the terms H_{pm} and H_{gu}:

$$H = H_e + H_N + H_{eN} \tag{1.15a}$$

with

$$H_e = -\frac{\hbar^2}{2m_e}\nabla_e^2 + \frac{e^2}{4\pi\varepsilon_0 |\vec{r}_1 - \vec{r}_2|} \tag{1.15b}$$

$$H_N = -\frac{\hbar^2}{2m_N}\nabla_{R_N}^2 + \frac{e^2}{4\pi\varepsilon_0 |\vec{R}_1 - \vec{R}_2|} \tag{1.15c}$$

$$H_{eN} = -\sum_{i=1}^{2}\sum_{k=1}^{2}\frac{e^2}{4\pi\varepsilon_0 |\vec{R}_i - \vec{r}_k|} \tag{1.15d}$$

To determine the molecule's energy levels, the Schrödinger eigenvalues equation must be solved for the molecular system. This equation involves the degrees of freedom of the nuclei and the electrons composing the molecule and, in general, there are no simple analytic solutions. It is therefore necessary to proceed in steps and the successive steps to resolve it are hereafter outlined.

1.4.2. *BO approximation*

The BO approximation is based on the ratio of the masses of an electron and a nucleus (as the mass ratio is such that $m_e/m_N \ll 1$; for example, of the order of 0.0005 for the hydrogen atom). It is considered that electrons instantaneously follow the nucleus' movement as a result of their lighter mass. The nuclei ($\vec{R}_N$) are fixed and the following eigenvalues equation is solved:

$$H\Phi_k(\vec{r}_e, \vec{R}_N) = (H_e + H_{eN})\Phi_k(\vec{r}_e, \vec{R}_N) = U_k(\vec{R}_N)\Phi_k(\vec{r}_e, \vec{R}_N) \tag{1.16}$$

where H is the Hamiltonian of the electronic system with $\vec{R}_N$ fixed.

We choose a solution in the form: $\Psi\left(\vec{r}_e,\vec{R}_N\right)=\sum_k \chi_k\left(\vec{R}_N\right)\Phi_k\left(\vec{r}_e,\vec{R}_N\right)$, to solve the equation:

$$\left(H_e + H_{eN} + H_N\right)\Psi\left(\vec{r}_e,\vec{R}_N\right) = E\Psi\left(\vec{r}_e,\vec{R}_N\right) \qquad [1.17]$$

Below, we will show how the BO approximation is applied to the simple diatomic molecule H_2. To simplify, we consider a solution in which the total wave function is a product of two functions only, $\Psi\left(\vec{r}_e,\vec{R}_N\right)=\Psi_e\Psi_N$. The calculations can easily be transposed to the function $\Psi\left(\vec{r}_e,\vec{R}_N\right)=\sum_k \chi_k\left(\vec{R}_N\right)\Phi_k\left(\vec{r}_e,\vec{R}_N\right)$.

For the system of electrons, we have:

$$\left(H_e + H_{eN}\right)\Psi_e = -\frac{\hbar^2}{2m_e}\nabla_e^2\Psi_e + \frac{e^2}{4\pi\varepsilon_0\left|\vec{r}_1-\vec{r}_2\right|}\Psi_e + H_{eN}\Psi_e = \varepsilon_e\Psi_e \qquad [1.18a]$$

where Ψ_e is a function of the electrons' coordinates and ε_e is an energy that is a constant for a given nuclear configuration. For another configuration, the solution Ψ_e, ε_e is different.

In this case, we can put equation [1.15c] in the following form:

$$H_N = -\frac{\hbar^2}{2m_N}\nabla_{R_N}^2 + \frac{e^2}{4\pi\varepsilon_0\left|\vec{R}_1-\vec{R}_2\right|} + \varepsilon_e\left(\left|\vec{R}_1 \quad \vec{R}_2\right|\right) \qquad [1.18b]$$

where the second term on the right corresponds to the Coulomb repulsion of the nuclei and the third term corresponds to the average value of the attraction exerted by the electronic cloud on the nuclei.

For the full system of electrons and nuclei, we have:

$$\left(H_e + H_{eN} + H_N\right)\Psi_e\Psi_N = E\Psi_e\Psi_N \qquad [1.19a]$$

and, in the developed form, the expression is:

$$E\Psi_e\Psi_N = -\Psi_N \frac{\hbar^2}{2m_e}\nabla_e^2\Psi_e + \frac{e^2}{4\pi\varepsilon_0|\vec{r}_1-\vec{r}_2|}\Psi_e\Psi_N + H_{eN}\Psi_e\Psi_N +$$
$$-\Psi_e\frac{\hbar^2}{2m_N}\nabla_{R_N}^2\Psi_N + \frac{e^2}{4\pi\varepsilon_0|\vec{R}_1-\vec{R}_2|}\Psi_e\Psi_N \qquad [1.19b]$$
$$-\Psi_N\frac{\hbar^2}{2m_N}\nabla_{R_N}^2\Psi_e + -2\frac{\hbar^2}{2m_N}\left(\frac{\partial\Psi_e}{\partial X_N}\frac{\partial\Psi_N}{\partial X_N} + \frac{\partial\Psi_e}{\partial Y_N}\frac{\partial\Psi_N}{\partial Y_N} + \frac{\partial\Psi_e}{\partial Z_N}\frac{\partial\Psi_N}{\partial Z_N}\right)$$

The BO approximation consists of neglecting the variations of the electronic wave function (first and second derivatives) with respect to the nuclear coordinates, which corresponds to a negligible influencc of the nuclei's movement on the electronic wave function. By taking the average of equation [1.19b] in the given electronic state, thereby:

$$E\langle\Psi_e|\Psi_e\rangle|\Psi_N\rangle = \langle\Psi_e|\frac{-\hbar^2}{2m_e}\nabla_e^2 + \frac{e^2}{4\pi\varepsilon_0|\vec{r}_1-\vec{r}_2|} + H_{eN}|\Psi_e\rangle|\Psi_N\rangle +$$
$$\langle\Psi_e|\Psi_e\rangle\left(-\frac{\hbar^2}{2m_N}\nabla_{R_N}^2 + \frac{e^2}{4\pi\varepsilon_0|\vec{R}_1-\vec{R}_2|}\right)|\Psi_N\rangle \qquad [1.20a]$$

then the eigenvalues equation of the nuclear terms can be written as:

$$E_N^e|\Psi_N^e\rangle = \left(-\frac{\hbar^2}{2m_N}\nabla_{R_N}^2 + \frac{e^2}{4\pi\varepsilon_0|\vec{R}_1-\vec{R}_2|} + \varepsilon_e\right)|\Psi_N^e\rangle$$
$$= \left(-\frac{\hbar^2}{2m_N}\nabla_{R_N}^2 + V_N^e\right)|\Psi_N^e\rangle \qquad [1.20b]$$

The nuclear wave functions only depend on the nuclear coordinates (vibration and rotation (section 2.4 of Chapter 2)) but are characterized by the electronic and nuclear quantum numbers. Similarly, the nuclear energy E_N^e and the potential V_N^e are evaluated with reference to the minimum of the potential curve of the electronic state considered (Figure 1.2). The molecule's total energy for a given electronic state is expressed as $E^e = V_N^{\min} + E_N^e$. In this expression, the first term only depends on the electronic quantum numbers (each state is labeled by the ordinate of the minimum of each potential curve and characterizes the molecule's electronic energy: Figure 1.2) and the second term depends on the electronic and nuclear quantum numbers.

1.4.3. *Adiabatic representation*

The total Hamiltonian is diagonal on the basis of exact eigenfunctions. An exact solution $\Psi_T^e(\vec{r}_e,\vec{R}_N)$ of the total Hamiltonian $H_T = H_e + H_{eN} + H_N$ must fulfill:

$$\left\langle \Psi_T^e \middle| H_T \middle| \Psi_T^{e'} \right\rangle = \delta_{ee'} E_T^e \qquad [1.21]$$

An exact solution can, in principle, be obtained under the form of an expansion on the basis of wave functions used in the BO approximation, such as:

$$\Psi_T^e\left(\vec{r}_e,\vec{R}_N\right) = \sum_{k,N_k} c_{e,N_k} \chi_{N_k}\left(\vec{R}_N\right) \Phi_k\left(\vec{r}_e,\vec{R}_N\right) \qquad [1.22]$$

The expansion coefficients are obtained by diagonalizing the total Hamiltonian matrix. The basis used is chosen such that the non-diagonal elements are as small as possible through the application of a unitary operation, such as with the Van Vleck method (section 2.2.6 of Chapter 2) [JOR 74, VAN 29]. When the BO approximation is valid, only a single expansion term subsists. When the non-diagonal elements are not negligible, only two expansion terms generally subsist.

1.4.4. *Diabatic representation*

When the non-diagonal elements are not negligible, particularly around a specific value of the internuclear distance (intersection of potential curves), the approximate wave functions used in the framework of the BO approximation introduce non-adiabatic coupling terms between the different electronic states of different symmetries. In general, the description of the molecular dynamics [LEF 04] of excited electronic states requires a description of the states beyond the BO approximation. In the adiabatic representation, the description is unique, but in the diabatic representation, this is not the case and there is an infinite number of possible representations. All the electronic states from any of these representations can form a complete basis set from the point of view of quantum mechanics. At the crossings, the adiabatic states can change dramatically, as a function of the nuclear coordinates. The derivatives with respect to the nuclei's displacements are very large and it can be very difficult to evaluate the non-adiabatic matrix terms. The non-diagonal elements $H_{ee'}$, which are larger than or of the same order of magnitude as the difference $H_{ee} - H_{e'e'}$ between the diagonal elements, are processed by diagonalizing the interaction block states. In the intersecting regions, the diabatic representation is more suitable. A processed example in the case of a diatomic molecule can be found, for example, in [LAN 75]. In a diatomic molecule, the adiabatic potential curves belonging to electronic states with the same symmetry generally do not cross.

1.5. Conclusion

This chapter introduces theoretical considerations focused on homonuclear and heteronuclear diatomic molecules. The full Hamiltonian operators in terms of all the degrees of freedom of a diatomic, both classical and the quantum ones, are recalled. The different scales of the different energies, eigenvalues of the Hamiltonian for the electronic, vibrational and rotational degrees of freedom, allow the total electronic and vibration–rotation state to be represented in a first approximation as a product of the Hamiltonian's eigenstates of each of the degrees of freedom within the framework of the BO approximation. From the symmetry properties of these eigenstates discussed, according to the nature of the molecule and its symmetry group, the interaction with the electromagnetic radiation that entails the matrix elements of an n-polar operator, leads to the spectral signature to be expected in the spectrum profile and to the selection rules. The symmetry of the fundamental electronic state and the type of diatomic molecule, whether homonuclear or heteronuclear, bring further modifications to the expected profile as a result of the alternation in intensity of the spectral lines when one takes the nucleus' spins into account. The case of the hydrogen molecule and its isotopologues is discussed to illustrate these considerations.

1.6. Appendix

A symmetry operation is an operation that brings an object into a new orientation which is equivalent to the old one. For a molecule, the symmetry operations with respect to its symmetry elements leave the molecule's equilibrium configuration invariant, meaning that the molecule coincides with itself. The possible symmetry elements of a molecule are the axes of rotation (proper axes), the inverse axes of rotations (improper axes), the planes of reflection symmetry and an inversion center. Five types of symmetry operations are depicted for a molecule:

– a rotation of an angle $2\pi/n$ around C_n, the rotation axis of the order n (if n = 1, this is the identity operation I); for a linear molecule, there is an infinite number of possible rotations of an angle φ around the internuclear axis z, which is labeled $C_z(\varphi)$ including the identity operation I;

– a rotation of an angle $2\pi/n$ around the axis C_n, followed by a reflection operation in a plane perpendicular to the axis which is termed S_n (improper rotation); for a linear molecule, there is an infinite number of improper rotations $S_z(\varphi)$ (rotation of an angle φ around the z axis followed from the reflection operation in a plane perpendicular to the axis passing through the center of the chemical bond), including the reflection operation with respect to the plane xy (rotation of an angle 0 around the z axis and an inversion S_2 or I with respect to the center of the chemical bond);

– for a symmetric linear molecule, there is an infinite number of rotations $C_t(\pi)$ of an angle π around the axes t perpendicular to z and passing through the origin;

– for a symmetric linear molecule, there is an infinite number of reflections σ_{zt} with respect to the planes zt containing the axis z and making any angle with the zx plane;

– the inversion with respect to the molecule's center of symmetry; for a symmetric linear molecule, the improper axis S_2 corresponds to the inversion with respect to the center of the chemical bond.

All the molecule's symmetry operations make up a symmetry point group in the mathematical sense [AMA 80a, CAM 88, LAN 75, WIL 80], and the symmetry elements may be grouped into categories. All of the symmetry operations O_j make up a group if the following conditions are fulfilled:

– the product $O_k = O_iO_j$ of two operations of the group is a member of the group;

– the product is associative $O_i(O_jO_k) = (O_iO_j)O_k$;

– there is an operation I called identity such that $IO_i = O_iI = O_i$;

– for each operation O_i, an operation O_i^{-1} called the inverse of O_i, can be defined, such that $O_iO_i^{-1} = O_i^{-1}O_i = I$.

If, in addition, we have $O_iO_j = O_jO_i$ (commutativity), the group is said to be Abelian.

The symmetry operations can be represented by unitary square matrices in a basis constructed on the molecule's equilibrium configuration and the set of matrices associated with each symmetry operation constitutes a representation of the group. The matrices have the same multiplication table as the symmetry point group.

So, a representation of dimension n can be built on the basis $\{\Psi_1, \Psi_2, \Psi_3, ..., \Psi_n\}$, composed of the set of square n × n matrices each associated with an operation of the group. Three cases may arise in the way these matrices are formed:

– if all the representation matrices comprise two blocks, respectively, with p and n-p lines and columns aligned on the main diagonal line, such as:

$$\begin{pmatrix} \overbrace{\begin{pmatrix} * & * \\ * & * \end{pmatrix}}^{p} & \begin{vmatrix} 0 & 0 & 0 \\ 0 & 0 & 0 \end{vmatrix} \\ \begin{vmatrix} 0 & 0 \\ 0 & 0 \\ 0 & 0 \end{vmatrix} & \underbrace{\begin{pmatrix} \bullet & \bullet & \bullet \\ \bullet & \bullet & \bullet \\ \bullet & \bullet & \bullet \end{pmatrix}}_{n-p} \end{pmatrix} \quad [1.23]$$

The representation appears under a reduced form. The first p components of the basis transform into themselves and constitute the basis of a representation of dimension p; the n-p other components of the basis also transform into themselves and constitute the basis of a representation of dimension n-p.

– a representation that does not appear under a reduced form is said to be reducible if it is possible to find a new basis whose components are n independent linear combinations of $\{\Psi_1, \Psi_2, \Psi_3,..., \Psi_n\}$ by a similarity operation and such that the representation constructed on the new basis $\{\Psi'_1, \Psi'_2, \Psi'_3,..., \Psi'_n\}$ appears under a reduced form. It is possible that the two new representations of dimensions p and n-p are themselves reducible;

– if, by a similarity operation, the order of the matrices cannot be reduced, the representation is said to be irreducible.

The number of irreducible representations of a group is equal to the number of classes in the group. An irreducible representation can be defined by the trace of the matrix (sum of the diagonal elements) also called the character of each matrix of the representation. In one same category, the characters are identical. When the group is Abelian, meaning that the symmetry operations commute, each operation constitutes one class by itself.

A group can be represented by a table in two dimensions with the classes of the symmetry operations in abscissa (for each column) and the irreducible representations in ordinate (for each line). The elements of the table are the characters of the matrices associated with the operations. These tables are called "Tables of Characters", in which we add a column comprised of the coordinates X, Y or Z and the rotations, J_X, J_Y or J_Z and the products that form a basis for the irreducible representation connected to the line. In the reference text book [WIL 80], one can find more detailed explanations on group theory's application to molecules.

As an example, for the group $C_{\infty v}$ (Table 1.2), we have the two-dimensional representation with the basis $\{\Psi_{-m}, \Psi_{+m}\}$ for $m \geq 1$. In this representation, the matrices associated with the rotations $C_z(\varphi)$ are in the form $\begin{pmatrix} e^{-im\varphi} & 0 \\ 0 & e^{+im\varphi} \end{pmatrix}$. This representation is not in a reduced form as the matrices associated with the reflections σ_{zt} do not have the same structure.

This representation is irreducible and two-dimensional and is of the symmetry type $\Pi, \Delta, \Phi, ...$ following the value of $m = 1, 2, 3, ...$. The trace Tr of the square 2 × 2 matrix, representing the rotation symmetry operation of angle φ around the z axis and built on the basis $\{\Psi_{-1}, \Psi_{+1}\}$ allows the symmetry type of the doubly degenerate states to be defined. This trace is expressed as $Tr = 2\cos m\varphi$ and according to the value of $m = 1, 2, 3, ..$, the states $\{\Psi_{-1}, \Psi_{+1}\}$ are of the symmetry type $\Pi, \Delta, \Phi, ...$.

2

Energy Levels of a Diatomic Molecule in Gaseous Phase

This chapter describes the theoretical model developed for studying the IR spectra, which result from the transitions between the vibration–rotation energy levels of diatomic molecules. The solution to the Schrödinger eigenvalues equation of the molecular system involves the degrees of freedom of the nuclei and electrons comprising the molecule and, in general, there are no simple analytical solutions. The Born–Oppenheimer approximation enables us to decouple the electrons' fast movement from that of the nuclei, which is much slower. For each electronic state, the nuclei's Hamiltonian allows the molecule's vibration–rotation movement to be studied. Decoupling from the translation movements is performed by studying the vibration–rotation movements in a frame bound to the molecule's equilibrium configuration (mobile frame) and whose origin coincides with its center of mass (Sayvetz–Eckart conditions). The vibration is then studied in the approximation of small amplitudes. In this case, a partial decoupling is achieved between the molecule's vibration and rotation movements. Van Vleck's contact method is used to split the Hamiltonian into blocks so that the molecules' IR spectra can be studied. The transitions between the energy levels that result from the interaction between the molecule's dipole moment or the induced dipole moment and the IR electromagnetic radiation lead, respectively, to absorption and/or emission spectra of the molecule or to the molecule's Raman spectra.

2.1. Introduction

An AB-type diatomic molecule is formed when the electrons localized on each of the two A and B atoms are shared. In general, there is a competition between the atoms' kinetic energy, which depends on the temperature of the ambient medium and tends to separate the atoms, and the potential electrical energy, which depends on the nuclear and electronic charges of the atoms and tends to merge them. This sharing is possible when the temperature is low enough to favor atom merging.

In such a structure, four types of movement can be distinguished:

– the electrons' movement around the nuclei;

– the nuclei's movement, comprising the molecule's overall translation and rotation, which is manifested under the form of a translation movement of the molecule's center of mass G and the rotation movement around it;

– the nuclei's vibration movement around their equilibrium position parallel to the bond's axis.

To describe these movements in space, it is required to know the molecule's set of electronic and nuclear coordinates. To do this, one can define a laboratory frame of reference equipped with a fixed frame $(O, \vec{X}, \vec{Y}, \vec{Z})$.

However, due to the fact that the electrons are much lighter than the nuclei (the mass ratio being such that $m_e/M_N << 1$; for example, of the order of 0.0005 for the hydrogen atom), their speeds are very high compared with those of the nuclei. Thus, when looking at the electrons' movement, it can be assumed that the latter are not affected, in a first approximation, by the nuclei's changes in movement status and vice versa. This enables the electrons' and nuclei's movements to be decoupled. This is the Born–Oppenheimer approximation. The total energy E of the molecule and the function Ψ describing its state can therefore be written as:

$$\begin{aligned} E &= E_e + E_N \\ \Psi &= \Psi_e \Psi_N \end{aligned} \qquad [2.1]$$

where (E_e, Ψ_e) characterizes the electronic movements given the positions of the nuclei, and (E_N, Ψ_N) characterizes the movements of the nuclei in a given electronic state.

Finally, as this study is dedicated to the isolated diatomic molecule's vibration–rotation spectra, this will be considered in its fundamental electronic state and its uniform translation movement will be ignored for the rest of this chapter. The translation movement is decoupled by studying the vibration–rotation movements in a frame (G, $\vec{X}$, $\vec{Y}$, $\vec{Z}$) bound to the molecule, a mobile frame whose origin coincides with its center of mass (Sayvetz–Eckart conditions).

Furthermore, the orders of magnitude of the vibration and rotation energies are very different, and because of the small amplitude of the vibration movement, it is possible to assume that the Born–Oppenheimer approximation is valid in the first instance to study these two movements separately.

The energy E_N and the function Ψ_N associated with the nuclear movements are therefore written as:

$$\begin{aligned} E_N &= E_{\text{vib}} + E_{\text{rot}} \\ \Psi_N &= \Psi_{\text{vib}} \Psi_{\text{rot}} \end{aligned} \qquad [2.2]$$

In the rest of this chapter, we will successively deal with:

– the pure vibration movement of a diatomic molecule;

– its rotation movement in its equilibrium configuration (rigid rotor);

– the vibration–rotation coupling.

Finally, we will determine the transition moments and the selection rules that allow the diatomic molecule's vibrational spectrum to be calculated.

2.2. Pure vibration movement of a diatomic molecule

In order to study the vibration movement of the diatomic molecule AB, corresponding to movement of small amplitudes, it is convenient to associate the molecule to a system located in its center of mass G with one of its axes $\vec{x}$, $\vec{y}$ or $\vec{z}$ parallel to the A-B bond (coinciding with $\vec{z}$, for example).

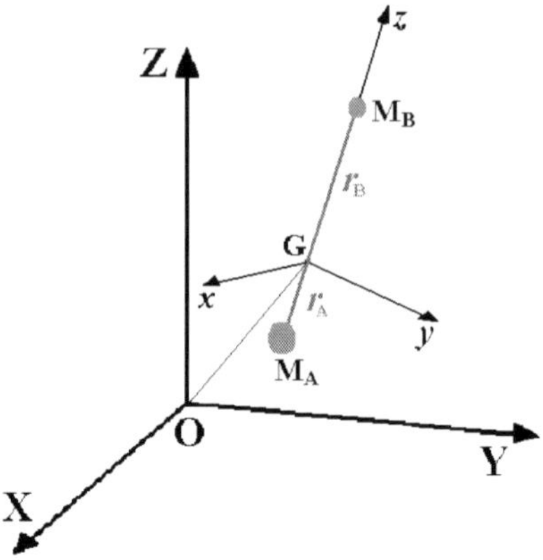

Figure 2.1. *Systems of axes bound to the laboratory and to the molecule's center of mass*

2.2.1. *Harmonic oscillator: classical processing*

When processed classically, the molecule can be modeled by a system of two masses M_A and M_B connected by a spring with a negligible mass, characterized by a real and positive force constant k and a length (interatomic distance) at equilibrium $r^e = r_A^e + r_B^e$, where r_A^e and r_B^e are the respective distances of M_A and M_B from the center of mass G defined by the relation:

$$M_A r_A^e = M_B r_B^e \tag{2.3}$$

with:

$$r_A^e = \frac{M_B}{(M_A + M_B)} r^e \text{ and } r_B^e = \frac{M_A}{(M_A + M_B)} r^e \tag{2.4}$$

The vibration movement of this system, said to be at one degree of freedom, is described by small oscillations in the vicinity of its stable equilibrium position.

During the vibration movement, at a given instant t, the masses M_A and M_B are found at the distances $r_A = r_A^e + \Delta r_A$ and $r_B = r_B^e + \Delta r_B$ from the center of mass G. Δr_A and Δr_B comprise the internal coordinates associated with the masses M_A and M_B (atoms A and B). The length of the spring has therefore been subjected to a variation $r - r^e = \Delta r_A + \Delta r_B = \Delta r$, the internal coordinate defining the AB molecule's strain.

The instantaneous speeds of M_A and M_B are, respectively, $d(\Delta r_A)/dt$ and $d(\Delta r_B)/dt$; the kinetic energy T_{vib} of the system is therefore given by:

$$T_{\text{vib}} = \frac{1}{2} M_A \left(\frac{d(\Delta r_A)}{dt} \right)^2 + \frac{1}{2} M_B \left(\frac{d(\Delta r_B)}{dt} \right)^2 = \frac{1}{2} \frac{M_A M_B}{(M_A + M_B)} \left(\frac{d(\Delta r)}{dt} \right)^2 \qquad [2.5]$$

This expression indicates that the system can be replaced by an equivalent model composed of a spring with a force constant k and with an equilibrium length r^e to the end of which a fictitious particle of mass μ is hung, while the other end is fixed. By putting $\Delta r = z$, this system's kinetic energy is expressed as:

$$T_{\text{vib}} = \frac{1}{2} \mu \left(\frac{dz}{dt} \right)^2 = \frac{1}{2} \mu . \dot{z}^2, \qquad [2.6]$$

where $\mu = \dfrac{M_A M_B}{M_A + M_B}$ is called "the reduced mass of the system" and $\dot{z}$ is the instantaneous speed of the fictitious particle's vibration movement (Figure 2.2).

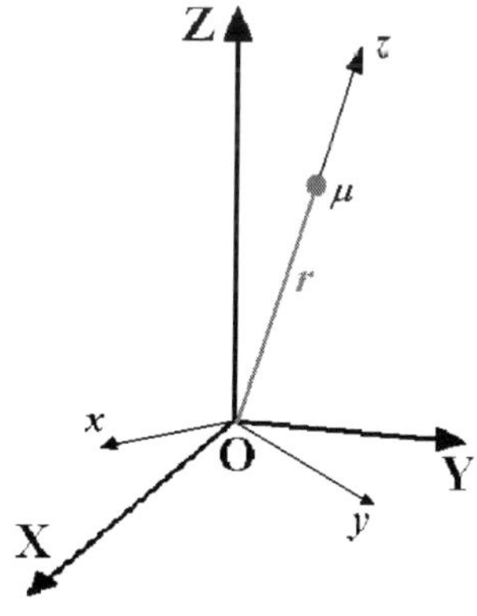

Figure 2.2. *Representation of the molecule's vibration movement by that of a fictitious particle mass μ*

Moreover, in the harmonic oscillator approximation (small movements of oscillations), the restoring force exerted by the spring on the mass μ is: $F = -kz$; it is derived from a scalar one-dimensional potential energy function, independent of the time variable that is expressed as:

$$V_{\text{vib}} = \frac{1}{2}kz^2 \qquad [2.7]$$

which has the profile of a parabolic curve with a minimum value $V = 0$ (arbitrary choice) for $r = r^e$, the stable equilibrium position. The force constant k therefore represents the second derivative of this function, calculated for $r = r^e$; its order of magnitude reflects the more or less harmonic property of the curve of the potential energy and therefore of the frequency and the amplitude of the vibrational movement. Morcover, we note that the first derivative of V_{vib} is zero for $r = r^e$.

Experience shows that simultaneous knowledge of the coordinate z and the speed $\dot{z}$ of the system determines its state and enables its subsequent movement to be predicted.

In non-relativistic classical mechanics, the movement's Lagrange function (called the "Lagrangian"), given by the difference between the kinetic energy T_{vib} and the potential energy V_{vib} of the "mass μ - spring" system, is expressed as:

$$L = T_{\text{vib}} - V_{\text{vib}} = \frac{1}{2}\mu.\dot{z}^2 - \frac{1}{2}k.z^2 \qquad [2.8]$$

Moreover, by virtue of the principle of least action (Hamilton's principle), this Lagrangian fulfills the equation:

$$\frac{d}{dt}\left(\frac{\partial L}{\partial \dot{z}}\right) - \frac{\partial L}{\partial z} = 0 \qquad [2.9]$$

called the "equation of motion". The conjugate momentum $p_z = \frac{\partial L}{\partial \dot{z}} = \frac{\partial T_{\text{vib}}}{\partial \dot{z}} = \mu\dot{z}$ of the coordinate z represents the linear momentum of the system's movement (impulsion).

Thus, the equation of the vibrational movement of the "mass μ-spring" system is written as:

$$\mu.\ddot{z} + k.z = 0 \qquad [2.10]$$

This is a differential equation of the second order whose solution is in the form:

$$z(t) = a\cos(\omega_0 t + \phi) \qquad [2.11]$$

where $\omega_0 = \sqrt{\frac{k}{\mu}}$ is the angular frequency, *a* and *φ*, respectively, represent the amplitude and the phase of the movement. These are two constants that can be determined from the initial conditions: $z(t=0) = z_0$ and $\dot{z}(t=0) = 0$. We therefore obtain:

$$z(t) = z_0 \cos(\omega_0 t) = z_0 \cos(2\pi\nu_0 t) \qquad [2.12]$$

In this equation, z_0 and $\nu_0 = \frac{\omega_0}{2\pi}$ (expressed in hertz) are, respectively, the amplitude and the frequency of the vibration movement of the harmonic oscillator.

Finally, the total energy of the moving system is given by:

$$E = \frac{1}{2}\mu\left(\dot{z}^2 + \omega_0^2 z^2\right) = \frac{1}{2\mu}p_z^2 + \frac{1}{2}\mu\omega_0^2 z^2 \qquad [2.13]$$

2.2.2. *Harmonic oscillator: quantum aspect*

When a molecule is subjected to an electromagnetic field, it absorbs or emits well-defined energy radiation (or frequency): a "photon". With a high-resolution spectroscope, the recorded absorption or emission spectrum is composed of well-separated lines. The molecule's vibrational motion is then said to be "quantified".

In quantum mechanics, the molecule's vibrational state, at a given instant, is completely determined by the vibrational wave function Ψ_{vib}, which determines its state at all subsequent instants. However, by assuming that the molecule is isolated, its wave function is independent of time and is only a function of the vibrational coordinate. The molecule's vibrational states are then said to be "stationary states". Their wave function Ψ_{v} and energy E_{v}, also called eigenfunction and eigenvalue, are determined by solving the Schrödinger equation:

$$H_{\text{vib}}\Psi_{\text{v}} = E_{\text{v}}\Psi_{\text{v}} \quad [2.14]$$

where v is a natural integer number called vibration quantum number and H_{vib} is the vibration Hamiltonian operator defined by:

$$H_{\text{vib}} = T_{\text{vib}} + V_{\text{vib}} \quad [2.15]$$

where T_{vib} and V_{vib} are the kinetic energy and potential energy operators, respectively.

In the harmonic approximation, and in order to quantum mechanically process the molecule's vibration movement, one introduces the dimensionless normal coordinate q and the dimensionless conjugate momentum p defined as:

$$q = \left(\frac{\mu\omega_0}{\hbar}\right)^{\frac{1}{2}} z \quad [2.16]$$

$$p = \left(\frac{1}{\hbar\mu\omega_0}\right)^{\frac{1}{2}} p_z = \left(\frac{\mu}{\hbar\omega_0}\right)^{\frac{1}{2}} \dot{z} \quad [2.17]$$

with $\hbar = h/2\pi$, where h is Planck's constant.

The vibration Hamiltonian operator, said to be of the order of zero, can be written in the form:

$$H_{\text{vib}}^0 = T_{\text{vib}} + V_{\text{vib}} = \frac{1}{2}\hbar\omega_0 p^2 + \frac{1}{2}\hbar\omega_0 q^2 \quad [2.18]$$

Furthermore, in quantum mechanics (the correspondence principle), the two classical conjugate variables q and p are replaced by the Hermitian operators q and $p=-i\frac{\partial}{\partial q}$ verifying the canonical commutation rule:

$$[q,p]=qp-pq=i \qquad [2.19]$$

These operators act on a space of quantum states with square integrable functions Ψ_{v}: "Hilbert space".

Finally, the Schrödinger equation associated with the one-dimensional Harmonic oscillator is given by:

$$H_{\text{vib}}^{0}\Psi_{\text{v}}^{0}=-\frac{1}{2}\hbar\omega_{0}\left(\frac{\partial^{2}}{\partial q^{2}}-q^{2}\right)\Psi_{\text{v}}^{0}=E_{\text{v}}^{0}\Psi_{\text{v}}^{0} \qquad [2.20]$$

Solving this equation enables the eigenenergies to be determined:

$$E_{\text{v}}^{0}=\left(\text{v}+\frac{1}{2}\right)\hbar\omega_{0}=\left(\text{v}+\frac{1}{2}\right)h\nu_{0} \qquad [2.21]$$

as well as the eigenfunctions:

$$\Psi_{\text{v}}^{0}(q)=N_{\text{v}}e^{-q^{2}/2}H_{\text{v}}(q) \qquad [2.22]$$

which are also noted as $|\Psi_{\text{v}}\rangle_{0}$ (ket) or even $|\text{v}\rangle_{0}$, comprising a full orthonormal set and with which the conjugate eigenfunctions $\Psi_{\text{v}}^{0*}(q)$ can be associated, which are noted as $\langle\Psi_{\text{v}}|$ (bra). In the previous expressions, ω_0 and ν_0 are the angular frequency (pulsation) and the frequency of the vibrational movement calculated classically, $N_{\text{v}}=\left(\frac{1}{2^{\text{v}}\,\text{v}!\sqrt{\pi}}\right)^{1/2}$ is the normalization coefficient of the eigenfunctions and $H_{\text{v}}(q)$ is the Hermite polynomial of the order v, defined by:

$$H_{v}(q)=(-1)^{v}e^{q^{2}}\frac{d^{v}}{dq^{v}}\left(e^{-q^{2}}\right) \quad [2.23]$$

The expressions of the first Hermite polynomials (v = 0,...,4) are given as:

$$H_{0}(q)=1 \quad [2.24a]$$

$$H_{1}(q)=2q \quad [2.24b]$$

$$H_{2}(q)=4q^{2}-2 \quad [2.24c]$$

$$H_{3}(q)=8q^{3}-12q \quad [2.24d]$$

$$H_{4}(q)=16q^{4}-48q^{2}+12 \quad [2.24e]$$

As can be observed, in the harmonic oscillator approximation, the eigenvalues E_{v}^{0} of the molecule's vibration comprise a set of discrete and equidistant energy levels, the energy difference between two consecutive levels is $\hbar\omega_{0}$. Moreover, the lowest energy level corresponding to the molecule's fundamental vibrational state is $E_{0}^{0}=\frac{1}{2}\hbar\omega_{0}$. This implies that even at the temperature 0 K, the molecule possesses an energy equal to $\frac{1}{2}\hbar\omega_{0}$.

Furthermore, the few examples of Hermite polynomials given above show that the wave function of each energy level has the same parity $(-1)^{v}$ as its vibrational quantum number.

In Figure 2.3, we show the profile of the eigenfunctions, as well as that of the harmonic potential energy and the energy levels, which are equidistant in the harmonic approximation.

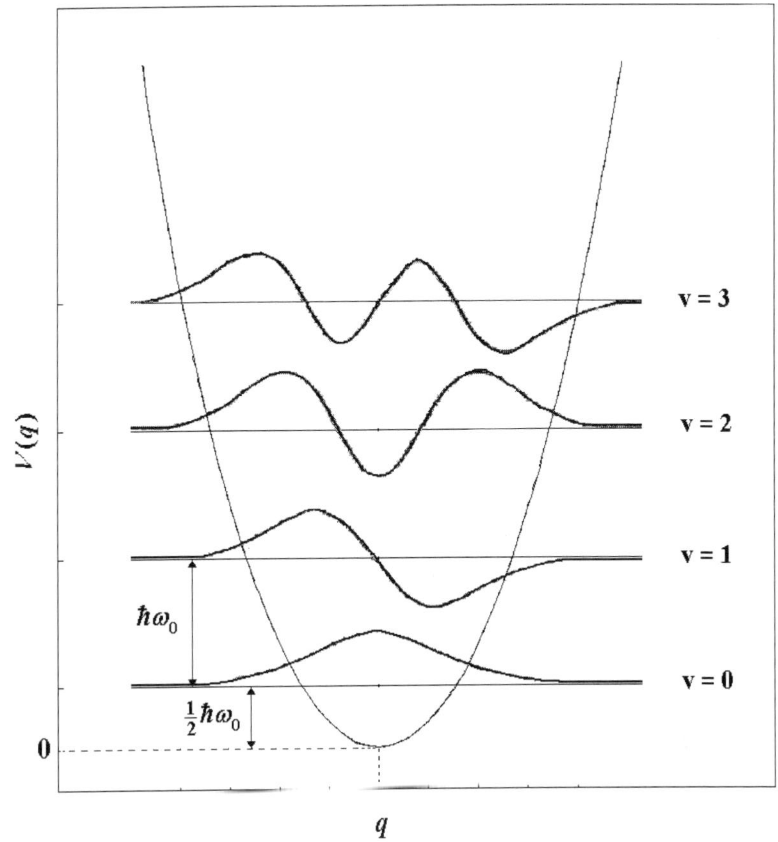

Figure 2.3. *Eigenfunctions, potential energy and energy levels*

NOTE.– In the spectroscopy domain, the frequency ν, the pulsation ω and energy are often expressed in wave numbers (cm^{-1}). We have $\nu\left(\text{cm}^{-1}\right)=\nu(\text{Hz})/c$, $\omega\left(\text{cm}^{-1}\right)=\omega(\text{Hz})/c$ and $E_{\text{v}}\left(\text{cm}^{-1}\right)=E_{\text{v}}(\text{joule})/c$, where c is the velocity of light.

2.2.3. *Transitions between two vibrational levels: selection rules*

When a molecule is subjected to electromagnetic radiation, there is an interaction between its electric dipole moment $\vec{\mu}$ and the electric field $\vec{E}$ of the radiation. If the molecule does not possess a permanent dipole moment, a dipole moment induced by the vibration may be present. The electrical field can also induce a dipole moment in the molecule such as $\vec{\mu} = \vec{\vec{\alpha}}.\vec{E}$, where $\vec{\vec{\alpha}}$ is the molecule's polarizability tensor. In any case, the interaction term is written in the form: $H' = -\vec{\mu}.\vec{E}$. Depending on the context of the dipole moment's existence, this term leads to the electrical dipole transitions (absorption or emission IR spectroscopy) or to the Raman transitions (radiation scattering).

When the molecule absorbs or emits a photon of energy $\hbar\omega_0$, its vibrational state undergoes a modification. It transits between its initial state of energy $E^0_{\mathrm{v}_i}$ and a final state of energy $E^0_{\mathrm{v}_f}$. This is manifested by the presence of a line in the near infrared spectrum and is characterized by small variations from its dipole moment $\vec{\mu}$, thus imposing constraints on the vibrational transition: "selection rules". The dipole transition moment ${}_0\langle \Psi_{\mathrm{v}_f} | \vec{\mu} | \Psi_{\mathrm{v}_i} \rangle_0 = {}_0\langle \mathrm{v}_f | \vec{\mu} | \mathrm{v}_i \rangle_0$ must be *non-zero*.

In order to determine these selection rules, it is possible to expand, in terms of the normal vibrational coordinate q, the molecule's dipole moment in the vicinity of its value at the equilibrium state $\vec{\mu}^e$ (permanent dipole moment that may be zero) in the mobile molecular frame:

$$\vec{\mu} = \vec{\mu}^e + \left(\frac{d\vec{\mu}}{dq}\right)_{q=0} q + \frac{1}{2!}\left(\frac{d^2\vec{\mu}}{dq^2}\right)_{q=0} q^2 + \frac{1}{3!}\left(\frac{d^3\vec{\mu}}{dq^3}\right)_{q=0} q^3 + \ldots, \qquad [2.25]$$

where the different terms of the expansion correspond to the dipole moment induced by the vibration and are generally truncated at the first order. The variations are assumed to be small.

In practice, the electrical dipole moment is expressed in the laboratory frame as the electric field of the electromagnetic radiation is directed along the fixed OZ axis of this frame. The component μ_Z of the dipole moment is connected to the components of the dipole moment on the molecular axes by the relation:

$$\mu_Z = \sum_{i=x,y,z} \varphi_{Zi}\mu_i \,, \qquad [2.26]$$

where φ_{Zi} are the directing cosines linking the fixed axis OZ to the molecular axes.

The moment of the vibrational transition leading to absorption or emission is written in the form:

$$_0\left\langle \Psi_{v_f} \middle| \vec{\mu} \middle| \Psi_{v_i} \right\rangle_0 = \int \Psi^{0*}_{v_f} \vec{\mu} \Psi^{0}_{v_i} dq \neq 0 \qquad [2.27]$$

or by replacing $\vec{\mu}$ with its expression and by using the "bra" and "ket" notations:

$$\begin{aligned} _0\left\langle v_f \middle| \vec{\mu} \middle| v_i \right\rangle_0 &= \vec{\mu}^e \, _0\left\langle v_f \middle| v_i \right\rangle_0 + \left(\frac{d\vec{\mu}}{dq}\right)_{q=0} \, _0\left\langle v_f \middle| q \middle| v_i \right\rangle_0 + \ldots \\ &= \vec{\mu}^e \delta_{v_i,v_f} + \left(\frac{d\vec{\mu}}{dq}\right)_{q=0} \sqrt{\frac{2^{v_f} v_f!}{2^{v_i} v_i!}} \left(\frac{1}{2}\delta_{v_i+1,v_f} + v_i \delta_{v_i-1,v_f}\right) + \ldots \end{aligned} \qquad [2.28]$$

where δ is the Kronecker symbol. In this expression, one term can be *non-zero*:

– if $\delta_{v_i,v_f} = 1$, this implies that $v_i = v_f$, so $\Delta v = 0$, no transition occurs;

– if $\delta_{v_i+1,v_f} = 1$, then $v_i + 1 = v_f$, with $\Delta v = +1$, which corresponds to an absorption transition;

– if $\delta_{v_i-1,v_f} = 1$, then $v_i - 1 = v_f$, with $\Delta v = -1$, which corresponds to an emission transition.

COMMENT.– If, in equation [2.25], one also accounts for the second-order term, the transitions corresponding to $\Delta v = \pm 2$ may occur.

The matrix elements of the operators q and p, which are useful for the following, are:

$$_0\left\langle v+1 \middle| q \middle| v \right\rangle_0 = {}_0\left\langle v \middle| q \middle| v+1 \right\rangle_0 = \sqrt{\frac{v+1}{2}} \qquad [2.29a]$$

$$_0\langle \mathrm{v}+1|p|\mathrm{v}\rangle_0 = -\,_0\langle \mathrm{v}|p|\mathrm{v}+1\rangle_0 = i\sqrt{\frac{\mathrm{v}+1}{2}} \quad [2.29b]$$

EXAMPLE 2.1.– The transition moment between the fundamental state $\mathrm{v}_i = 0$ and the first excited state $\mathrm{v}_f = 1$ is:

$$_0\langle 1|\vec{\mu}|0\rangle_0 = \left(\frac{d\vec{\mu}}{dq}\right)_{q=0} \frac{\sqrt{2}}{2}$$

2.2.4. "Creation" and "annihilation" operators

For practical reasons, the processing of the real oscillator is based on the harmonic oscillator basis, by introducing the concept of "$\mathrm{a}^\dagger$ creation" and "a annihilation" (second quantification), hermitic and conjugate operators, defined as:

$$\mathrm{a}^\dagger = \frac{1}{\sqrt{2}}(q - ip) \quad [2.30]$$

$$\mathrm{a} = \frac{1}{\sqrt{2}}(q + ip) \quad [2.31]$$

verifying the commutation rule:

$$\left[\mathrm{a}, \mathrm{a}^\dagger\right] = \mathrm{a}\mathrm{a}^\dagger - \mathrm{a}^\dagger\mathrm{a} = 1 \quad [2.32]$$

Inversely, the q and p operators are written as:

$$q = \frac{1}{\sqrt{2}}\left(\mathrm{a}^\dagger + \mathrm{a}\right) \quad [2.33]$$

$$p = \frac{i}{\sqrt{2}}\left(\mathrm{a}^\dagger - \mathrm{a}\right) \quad [2.34]$$

The matrix elements of the "$a^\dagger$ creation" and "a annihilation" operators are given by:

$$_0\langle v+1|a^\dagger|v\rangle_0 = \sqrt{v+1} \quad [2.35]$$

$$_0\langle v|a|v+1\rangle_0 = \sqrt{v+1} \quad [2.36]$$

and their application to any eigenvector $|v\rangle_0$ induces, respectively, the "creation" and the "annihilation" of a vibration photon, such as:

$$a^\dagger|v\rangle_0 = \sqrt{v+1}|v+1\rangle_0 \quad [2.37]$$

$$a|v\rangle_0 = \sqrt{v}|v-1\rangle_0 \quad [2.38]$$

with

$$a|0\rangle_0 = 0 \quad [2.39]$$

In light of this, the Hamiltonian operator H^0_{vib} of the Harmonic oscillator is written as:

$$H^0_{vib} = \frac{1}{2}\hbar\omega_0\left(p^2+q^2\right) = \hbar\omega_0\left(a^\dagger a+\frac{1}{2}\right) \quad [2.40]$$

Finally, it is easy to show that:

$$H^0_{vib}|v\rangle_0 = E^0_v|v\rangle_0 \Rightarrow \hbar\omega_0\left(a^\dagger a+\frac{1}{2}\right)|v\rangle_0 = \hbar\omega_0\left(v+\frac{1}{2}\right)|v\rangle_0 \quad [2.41]$$

From this expression, the action of the hermitic operator N = $a^\dagger a$ on an eigenvector $|v\rangle_0$ leads to the multiplication of this eigenvector by the quantum number v, such that:

$$N|v\rangle_0 = a^\dagger a|v\rangle_0 = v|v\rangle_0 \quad [2.42]$$

This property is very useful in the perturbative calculations, which are applied to the anharmonic oscillator.

2.2.5. *Anharmonic oscillator*

In the harmonic oscillator model, it is clear that an infinite potential energy would be needed to break the diatomic molecule's A-B bond. Yet the physical reality is completely different; the energy necessary to break apart the two atoms from each other remains finite and tends toward a value called "dissociation energy". Moreover, the observations in spectroscopy show that the spacing between the successive energy levels decreases when the vibrational quantum number increases.

A more realistic model is that of the anharmonic oscillator (Figure 2.4).

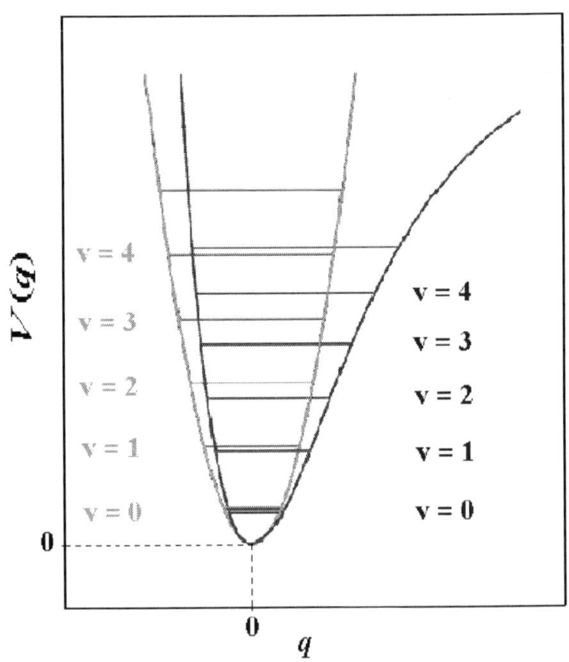

Figure 2.4. *Potential energy of a diatomic molecule modeled like an anharmonic oscillator (blue curve). For a color version of this figure, see www.iste.co.uk/dahoo/infrared.zip*

In this model, the expansion series of the potential energy function must be pursued in the higher orders in the form:

$$\begin{aligned} V_{\text{vib}} &= \frac{1}{2}\hbar\omega_0 q^2 + k_3 q^3 + k_4 q^4 + k_5 q^5 + \ldots \\ &= \frac{1}{2}\hbar\omega_0 q^2 + \Delta V_{\text{vib}}^{\text{anhar}}(q) \end{aligned} \qquad [2.43]$$

and the Hamiltonian is written as:

$$H_{\text{vib}}^{\text{anhar}} = H_{\text{vib}}^{0} + k_3 q^3 + k_4 q^4 + k_5 q^5 + \ldots, \qquad [2.44]$$

where the coefficients k_3, k_4, k_5, …, are constants associated with the anharmonicity of the vibration movement and represent the successive derivatives (third, fourth, fifth, etc.) of the potential energy function with respect to the vibrational normal coordinate q. We note that these anharmonicity terms are much smaller than the harmonic term and are therefore processed as perturbations.

To determine the eigenenergy E_{v} and the eigenvector $|\text{v}\rangle$ associated with the vibration level v of the anharmonic oscillator, we use the stationary perturbations theory such that:

$$\begin{aligned} E_{\text{v}} &= E_{\text{v}}^{0} + {}_0\langle \text{v}| k_3 q^3 + k_4 q^4 + k_5 q^5 + \ldots |\text{v}\rangle_0 \\ &+ \sum_{\text{v}' \neq \text{v}} \frac{\left| {}_0\langle \text{v}'| k_3 q^3 + k_4 q^4 + k_5 q^5 + \ldots |\text{v}\rangle_0 \right|^2}{E_{\text{v}}^{0} - E_{\text{v}'}^{0}} + \ldots \end{aligned} \qquad [2.45]$$

and

$$|\text{v}\rangle = |\text{v}\rangle_0 + \sum_{\text{v}' \neq \text{v}} \frac{{}_0\langle \text{v}'| k_3 q^3 + k_4 q^4 + k_5 q^5 + \ldots |\text{v}\rangle_0}{E_{\text{v}}^{0} - E_{\text{v}'}^{0}} |\text{v}'\rangle_0 + \ldots \qquad [2.46]$$

with ${}_0\langle \text{v}| q^{2\text{n}+1} |\text{v}\rangle_0 = 0$.

In the first perturbation order, the correction at the eigenvector $|\text{v}\rangle$ is a linear superposition of all the non-perturbed eigenvectors other than $|\text{v}\rangle_0$: it is said that the perturbation terms cause the state $|\mathrm{v}\rangle_0$ to be contaminated by the other eigenstates of H^0_{vib}.

We limit the expansion of the potential energy function V_{vib} to the fifth order. The elements ${}_0\langle \mathrm{v'}|q^{\mathrm{n}}|\mathrm{v}\rangle_0$ are calculated by replacing the terms q^{n} expressed in terms of the $\mathbf{a}^\dagger$ creation and $\mathbf{a}$ annihilation operators and by using the expressions given in the Appendix (section 2.5).

The energy correction in the second perturbation order enables the level v eigenenergy to be determined as:

$$E_{\mathrm{v}} = \hbar\omega_0 C_1\left(\mathrm{v}+\frac{1}{2}\right)+C_2\left(\mathrm{v}+\frac{1}{2}\right)^2+C_3\left(\mathrm{v}+\frac{1}{2}\right)^3+C_4\left(\mathrm{v}+\frac{1}{2}\right)^4+C_0, \qquad [2.47]$$

where C_1 is a dimensionless constant, C_2, C_3, C_4 and C_0 are constants that have the dimension of an energy. They are given by:

$$C_1 = 1-\frac{1}{16(\hbar\omega_0)^2}\left(67k_4^2+190k_3k_5\right) \qquad [2.48a]$$

$$C_2 = \frac{3}{2}k_4-\frac{5}{32\hbar\omega_0}\left(24k_3^2+217k_5^2\right) \qquad [2.48b]$$

$$C_3 = -\frac{1}{4\hbar\omega_0}\left(17k_4^2+70k_3k_5\right) \qquad [2.48c]$$

$$C_4 = -\frac{315}{16\hbar\omega_0}k_5^2 \qquad [2.48d]$$

$$C_0 = \frac{3}{8}k_4-\frac{1}{256\hbar\omega_0}\left(112k_3^2+1107k_5^2\right) \qquad [2.48e]$$

The eigenvector correction at the first perturbation order leads to:

$$\begin{aligned}|\mathrm{v}\rangle = {}&|\mathrm{v}\rangle_0 + A_{(-1)}|\mathrm{v}\text{-}1\rangle_0 + A_{(+1)}|\mathrm{v}+1\rangle_0 + A_{(-2)}|\mathrm{v}\text{-}2\rangle_0 + A_{(+2)}|\mathrm{v}+2\rangle_0 + A_{(-3)}|\mathrm{v}\text{-}3\rangle_0\\ &+ A_{(+3)}|\mathrm{v}+3\rangle_0 + A_{(-4)}|\mathrm{v}\text{-}4\rangle_0 + A_{(+4)}|\mathrm{v}+4\rangle_0 + A_{(-5)}|\mathrm{v}\text{-}5\rangle_0 + A_{(+5)}|\mathrm{v}+5\rangle_0\end{aligned} \qquad [2.49]$$

with

$$A_{(-1)} = \frac{1}{4\hbar\omega_0}\sqrt{\frac{\text{v}}{2}}\left(6\text{v}k_3 + 5\left(2\text{v}^2+1\right)k_5\right) \quad [2.50a]$$

$$A_{(+1)} = -\frac{1}{4\hbar\omega_0}\sqrt{\frac{\text{v}+1}{2}}\left(6\left(\text{v}+1\right)k_3 + 5\left(2\text{v}^2+4\text{v}+3\right)k_5\right) \quad [2.50b]$$

$$A_{(-2)} = \frac{\left(2\text{v}-1\right)}{4\hbar\omega_0}\sqrt{\text{v}\left(\text{v-1}\right)}k_4 \quad [2.50c]$$

$$A_{(+2)} = -\frac{\left(2\text{v}+3\right)}{4\hbar\omega_0}\sqrt{\left(\text{v}+1\right)\left(\text{v}+2\right)}k_4 \quad [2.50d]$$

$$A_{(-3)} = \frac{1}{12\hbar\omega_0}\sqrt{\frac{\text{v}\left(\text{v}-1\right)\left(\text{v}-2\right)}{2}}\left(2k_3 + 5\left(\text{v}-1\right)k_5\right) \quad [2.50e]$$

$$A_{(+3)} = -\frac{1}{12\hbar\omega_0}\sqrt{\frac{\left(\text{v}+1\right)\left(\text{v}+2\right)\left(\text{v}+3\right)}{2}}\left(2k_3 + 5\left(\text{v}+2\right)k_5\right) \quad [2.50f]$$

$$A_{(-4)} = \frac{1}{16\hbar\omega_0}\sqrt{\text{v}\left(\text{v-1}\right)\left(\text{v-2}\right)\left(\text{v-3}\right)}k_4 \quad [2.50g]$$

$$A_{(+4)} = -\frac{1}{16\hbar\omega_0}\sqrt{\left(\text{v}+1\right)\left(\text{v}+2\right)\left(\text{v}+3\right)\left(\text{v}+4\right)}k_4 \quad [2.50h]$$

$$A_{(-5)} = \frac{1}{20\hbar\omega_0}\sqrt{\frac{\text{v}\left(\text{v}-1\right)\left(\text{v}-2\right)\left(\text{v}-3\right)\left(\text{v}-4\right)}{2}}k_5 \quad [2.50i]$$

$$A_{(+5)} = -\frac{1}{20\hbar\omega_0}\sqrt{\frac{\left(\text{v}+1\right)\left(\text{v}+2\right)\left(\text{v}+3\right)\left(\text{v}+4\right)\left(\text{v}+5\right)}{2}}k_5 \quad [2.50j]$$

The probability density that the anharmonic oscillator lies in the state $|\mathrm{v}\rangle$ can easily be obtained as:

$$\begin{aligned}\langle \mathrm{v}|\mathrm{v}\rangle = 1 &+ \left(\tfrac{29}{24}k_3^2 + \tfrac{39}{32}k_4^2 + \tfrac{155}{24}k_3k_5 + \tfrac{4447}{480}k_5^2\right) + \left(\tfrac{32}{9}k_3^2 + \tfrac{211}{64}k_4^2 + \tfrac{185}{9}k_3k_5 + \tfrac{118841}{3600}k_5^2\right)\mathrm{v} \\ &+ \left(\tfrac{41}{12}k_3^2 + \tfrac{487}{128}k_4^2 + \tfrac{1015}{36}k_3k_5 + \tfrac{6719}{144}k_5^2\right)\mathrm{v}^2 + \left(\tfrac{41}{18}k_3^2 + \tfrac{65}{64}k_4^2 + \tfrac{275}{18}k_3k_5 + \tfrac{2509}{60}k_5^2\right)\mathrm{v}^3 \\ &+ \left(\tfrac{65}{128}k_4^2 + \tfrac{275}{36}k_3k_5 + \tfrac{11567}{720}k_5^2\right)\mathrm{v}^4 + \left(\tfrac{11567}{1800}k_5^2\right)\mathrm{v}^5\end{aligned} \quad [2.51]$$

2.2.6. *Contact transformation method*

The contact transformation method consists of using the harmonic oscillator's eigenfunctions, the zeroth-order Hamiltonian H_{vib}^0 of equation [2.44], so as to calculate the corrections to be made to the eigenvalues of this Hamiltonian when one includes the anharmonic part ($k_3q^3 + k_4q^4 + k_5q^5 + \ldots$ of the vibrational potential) that drives the movement of the nuclei.

The Hamiltonian is generally expressed in the following form:

$$H_{\mathrm{vib}}^{\mathrm{anhar}} = H_{\mathrm{vib}}^0 + H^{(1)} + H^{(2)} + H^{(3)} + \ldots, \quad [2.52]$$

where $H^{(1)} = k_3q^3 = k_{111}q^3$, $H^{(2)} = k_4q^4 = k_{1111}q^4$ and $H^{(3)} = k_5q^5 = k_{11111}q^5$, and the unitary transformations are used in the form of an exponential operator $T = e^{iS}$ as introduced by Van Vleck [VAN 29] and further detailed by Jordahl [JOR 34] and Kemble [KEM 37].

The contact transformation method is applied by expressing the Schrödinger equation:

$$H_{\mathrm{vib}}^0 |\Psi_{\mathrm{v}}\rangle_0 = E_{\mathrm{v}}^0 |\Psi_{\mathrm{v}}\rangle_0 \quad [2.53]$$

in the following transformed form:

$$\tilde{H}_{\mathrm{vib}}^0 |\Phi_{\mathrm{v}}\rangle_0 = E_{\mathrm{v}}^0 |\Phi_{\mathrm{v}}\rangle_0 \quad [2.54]$$

such as:

$$\tilde{H}^{0}_{\text{vib}} = e^{iS} H^{0}_{\text{vib}} e^{-iS} \qquad [2.55]$$

and

$$|\Phi_{\text{v}}\rangle_0 = e^{iS} |\Psi_{\text{v}}\rangle_0 \qquad [2.56]$$

where S is a Hermitic and antisymmetric operator with respect to the time reversal such that e^{iS} is a unitary and invariant operator in a time-reversal symmetry operation.

If the transformed Hamiltonian $\tilde{H}^{0}_{\text{vib}}$ is diagonal in the basis $|\Phi_{\text{v}}\rangle_0$, the diagonal elements are the eigenvalues E of the initial Hamiltonian H^{0}_{vib} and the eigenfunctions are $|\Psi_{\text{v}}\rangle_0 = e^{-iS} |\Phi_{\text{v}}\rangle_0$. More generally, $\tilde{H}^{0}_{\text{vib}}$ can be diagonal by blocks. In this case, it is necessary to transform the Hamiltonian $\tilde{H}^{0}_{\text{vib}}$ again by using a unitary operator to diagonalize each block.

The method has been generalized in molecular physics to study the spectroscopy of molecules [ALI 85, AMA 57a, AMA 57b, AMA 58, AMA 71, BAR 61, BRO 03, BUN 98, CAM 85, GOL 56, JOR 74a, JOR 74b, MIL 72, NIE 51, PAP 82, SCH 39] and consists of using the eigenfunctions of the harmonic oscillator corresponding to the zeroth-order Hamiltonian of equation [2.44] to calculate the corrections to be made to the eigenvalues of this Hamiltonian when the anharmonic part of the electronic potential is included. The unitary operator $T = e^{iS}$ is applied to the Hamiltonian $H^{\text{anhar}}_{\text{vib}} = H^{0}_{\text{vib}} + H^{(1)} + H^{(2)} + H^{(3)} + \ldots$ to construct effective Hamiltonians at different perturbation orders so that they are diagonal by blocks in the eigenfunctions basis of the zeroth-order Hamiltonian H^{0}_{vib}. This method is generally applied for studying not only the molecules' vibration–rotation Hamiltonian that are considered to be rigid or semirigid, but also the non-rigid (floppy) molecules [SIB 88, SUG 99] and termed CVPT (Canonical Van Vleck Perturbation Theory). With regard to diatomic molecules, a very good description for using the contact transformation leading to an effective Hamiltonian can be found in [GEE 79, YAO 76] and more specifically for the diatomic molecules in fundamental and excited electronic states in [LEF 04].

To apply the method, the Hamiltonian is first expressed in the form of a limited expansion in λ, a dimensionless parameter having a value lower than 1, which enables the orders of magnitude of the expression to be quantified in terms of λ, such as:

$$H_{\text{vib}}^{\text{anhar}} = H_0 + \lambda H_1 + \lambda^2 H_2 + \lambda^3 H_3 + \ldots \qquad [2.57]$$

Then, the correction to be made to the zeroth-order energies in the first order in λ is evaluated by applying a first contact transformation $T_1 = e^{i\lambda S_1}$ to H_{vib} so as to construct the effective Hamiltonian in the first order:

$$H'_{\text{vib}} = T_1 H_{\text{vib}}^{\text{anhar}} T_1^{-1} = H'_0 + \lambda H'_1 + \lambda^2 H'_2 + \ldots \qquad [2.58]$$

From the expansion of the unitary exponential operator:

$$exp(\pm i\lambda S_1) = 1 \pm i\lambda S_1 - \frac{\lambda^2}{2} S_1^2 + \ldots \qquad [2.59]$$

and from the expansion of the second term of equation [2.58], the following set of equalities or identities can be established:

$$H'_0 = H_0 \qquad [2.60a]$$

$$H'_1 = H_1 + [iS_1, H_0] = H_1 + iS_1 H_0 - iH_0 S_1 \qquad [2.60b]$$

$$H'_2 = H_2 + [iS_1, H_1] + \frac{1}{2}[iS_1, [iS_1, H_0]] \qquad [2.60c]$$

In a more compact form, the result of the contact transformation can be written as:

$$H' = H + \sum_{n=1}^{\infty} \frac{(i\lambda)^n}{n!} [S_1, H]^{(n)} \quad [2.61]$$

in which

$$[S_1, H]^{(n)} = \left[S_1, \left[S_1 ..., [S_1, H]...\right]\right] \quad [2.62]$$

We determine S_1 so that H_1' is factored in blocks in the eigenfunctions' basis of H_0.

To construct H'', the effective Hamiltonian in the second order, one applies $T_2 = e^{i\lambda^2 S_2}$ to H' so that:

$$H'' = T_2 H' T_2^{-1} = T_2 T_1 H T_1^{-1} T_2^{-1} = H_0'' + \lambda H_1'' + \lambda^2 H_2'' + ... \quad [2.63a]$$

with

$$H_0'' = H_0' = H_0 \quad [2.63b]$$

$$H_1'' = H_1' \quad [2.63c]$$

$$H_2'' = H_2' + [iS_2, H_0] \quad [2.63d]$$

In a more compact form, the result of the second contact transformation can be written as:

$$H'' = H' + \sum_{n=1}^{\infty} \frac{(i\lambda^2)^n}{n!} \left[S_2, H'\right]^{(n)} \quad [2.64]$$

We determine S_2 so that H_2'' is factored in blocks in the eigenfunctions' basis of H_0.

Thus, we calculate S_1 from:

$$\langle v^{'}|H_1|v^{"}\rangle + i\langle v^{'}|S_1|v^{"}\rangle(E_{v^{'}} - E_{v^{"}}) = 0 \qquad [2.65]$$

and S_2 from:

$$\langle v^{'}|H_2^{'}|v^{"}\rangle + i\langle v^{'}|S_2|v^{"}\rangle(E_{v^{'}} - E_{v^{"}}) = 0, \qquad [2.66]$$

where $|v^{"}\rangle$ and $|v^{'}\rangle$ are eigenfunctions of the Hamiltonian H_0.

After n contact transformations, for a given order n, the calculation leads to the following effective Hamiltonian:

$$H^{(n)} = H_0^{(0)} + \lambda H_1^{(1)} + \lambda^2 H_2^{(2)} + \ldots + \lambda^n H_n^{(n)} + \ldots.., \qquad [2.67]$$

where all the terms up to the order n, ($H_n^{(n)}$), are factorized in the form of blocks in the eigenfunctions basis of H_0. The equivalences $H_0^{(0)} \equiv H_0; H_1^{(1)} \equiv H_1^{'}; H_2^{(2)} \equiv H_2^{"}$; etc. have been used in this expression.

In the case of a diatomic molecule, the operators S_1 and S_2 are given by:

$$S_1 = -\frac{k_{111}}{\omega_0}\left(\frac{2}{3}p^3 + qpq\right) \qquad [2.68]$$

$$\begin{aligned} S_2 = &\frac{1}{32}\left(\frac{k_{111}}{\omega_0}\right)^2 \left\{15\left(p^3q + qp^3\right) + 9\left(q^3p + pq^3\right)\right\} \\ &- \frac{1}{32}\left(\frac{k_{1111}}{\omega_0}\right)\left\{6\left(p^3q + qp^3\right) + 10\left(q^3p + pq^3\right)\right\} \end{aligned} \qquad [2.69]$$

From S_1, for the energy in the first order (using the matrix elements from the appendices A2.2 and A2.3), we find:

$$E_v^{'} = \hbar\omega_0\left(v + \frac{1}{2}\right) + \frac{3}{2}k_{1111}\left(\left(v + \frac{1}{2}\right)^2 + \frac{1}{4}\right) - \left(\frac{k_{111}^2}{8\hbar\omega_0}\right)\left(30v^2 + 30v + 11\right) \qquad [2.70]$$

Applying S_2 shows that the energy correction in the second order is zero so that S_1 leads directly to an energy correction of order 2.

To calculate the transition moment that leads to absorption or emission (equations [2.27] and [2.28]), it is necessary to use the transformed wave functions corresponding to the correction order (equation [2.57]). If two contact transformations are applied, the matrix elements of the dipole operator are calculated as follows:

$$_0\left\langle \Psi_{v_f} \left| T_2 T_1 \vec{\mu} T_1^{-1} T_2^{-1} \right| \Psi_{v_i} \right\rangle_0 = \int \Psi_{v_f}^{0*} T_2 T_1 \vec{\mu} T_1^{-1} T_2^{-1} \Psi_{v_i}^0 \, dq \neq 0 \qquad [2.71]$$

In the molecular frame, the dipole moment (equation [2.25]) can be expressed in the form of a limited expansion in λ. For a diatomic molecule, the dipole moment is directed along the internuclear axis Gz, and it is written as:

$$\begin{aligned} \mu_z &= \mu_z^e + \mu_z^{(1)} q + \lambda \frac{1}{2!} \mu_z^{(2)} q^2 + \lambda^2 \frac{1}{3!} \mu_z^{(3)} q^3 + \ldots + \lambda^n \frac{1}{(n+1)!} \mu_z^{(n+1)} q^{n+1} \\ &= \mu_{0z} + \lambda \mu_{1z} + \lambda^2 \mu_{2z} + \ldots + \lambda^n \mu_{nz} \end{aligned} \qquad [2.72]$$

In this development, we put $\mu_z^{(n)} = \left(\frac{d^n \mu_z}{dq^n} \right)_{q-0}$ and in the order n we find the terms of vibrational degree (n+1) with respect to the normal coordinate.

In the molecular frame, the transformed dipole moment is expressed as:

$$\mu_z'' = T_2 \mu_z' T_2^{-1} - T_2 T_1 \mu_z T_1^{-1} T_2^{-1} = \mu_{0z}'' + \lambda \mu_{1z}'' + \lambda^2 \mu_{2z}'' + \ldots \qquad [2.73a]$$

with

$$\mu_{0z}'' = \mu_{0z}' = \mu_{0z} \qquad [2.73b]$$

$$\mu_{1z}'' = \mu_{1z}' = \mu_{1z} + [iS_1, \mu_{0z}] \qquad [2.73c]$$

$$\mu_{2z}'' = \mu_{2z}' + [iS_2, \mu_{0z}] = \mu_{2z} + [iS_1, \mu_{1z}] + \frac{1}{2}[iS_1, [iS_1, \mu_{0z}]] + [iS_2, \mu_{0z}] \qquad [2.73d]$$

which is written in the following condensed form:

$$\mu_z^{''} = \mu_z + [iS_1, \mu_z] + \frac{1}{2}[iS_1, [iS_1, \mu_z]] + [iS_2, \mu_z] \quad [2.74]$$

In the fixed frame, the transformed dipole moment is expressed as:

$$\mu_Z^{''} = \mu_Z + [iS_1, \mu_Z] + \frac{1}{2}[iS_1, [iS_1, \mu_Z]] + [iS_2, \mu_Z] \quad [2.75]$$

which is written according to the direction cosines in the form:

$$\mu_Z^{''} = \frac{1}{2}\sum_{i=x,y,z}\left(\phi_{Zi}^{''}\mu_i^{''} + \mu_i^{''}\phi_{Zi}^{''}\right) \quad [2.76]$$

with:

$$\varphi_{Zi}^{''} = \varphi_{Zi} + [iS_1, \phi_{Zi}] + \frac{1}{2}[iS_1, [iS_1, \phi_{Zi}]] + [iS_2, \phi_{Zi}] \quad [2.77]$$

Finally, the first three vibrational transition moments are expressed as:

$$\langle \mathrm{v}|\mu_z|\mathrm{v}\rangle = \mu_z^e + \left(\mathrm{v}+\frac{1}{2}\right)\left\{\frac{1}{2}\mu_z^{(2)} - 3\mu_z^{(1)}\left(\frac{k_{111}}{\omega_0}\right)\right\} \quad [2.78a]$$

$$\langle \mathrm{v}|\mu_z|\mathrm{v}+1\rangle = \sqrt{\frac{\left(\mathrm{v}+\frac{1}{2}\right)}{2}}\mu_z^{(1)}\Theta \quad [2.78b]$$

in which

$$\Theta = \left\{1 + (\mathrm{v}+1)\left[\frac{1}{4}\frac{\left(\mu_z^{(3)}\right)}{\left(\mu_z^{(1)}\right)} - \frac{5}{2}\frac{\left(\mu_z^{(2)}\right)}{\left(\mu_z^{(1)}\right)}\left(\frac{k_{111}}{\omega_0}\right) + \frac{11}{4}\left(\frac{k_{111}}{\omega_0}\right)^2 - \frac{3}{2}\left(\frac{k_{1111}}{\omega_0}\right)\right]\right\}$$

$$\langle \mathrm{v}|\mu_z|\mathrm{v}+2\rangle = \sqrt{\frac{(\mathrm{v}+1)(\mathrm{v}+2)}{16}}\left\{\mu_z^{(2)} + 2\mu_z^{(1)}\left(\frac{k_{111}}{\omega_0}\right)\right\} \quad [2.78c]$$

2.3. Rotation movement of a rigid diatomic molecule

When the molecule's two nuclei are fixed in their equilibrium position on the axis $\vec{z}$ of the bond, the molecule is therefore assumed to be rigid: $r = r^e$ or $q = 0$ and the vibration potential energy is zero. The rotation movement of the free molecule (the translation movement is ignored) can be described in the frame (G, $\vec{X}$, $\vec{Y}$, $\vec{Z}$) by its polar angles φ (precession) and θ (nutation). However, similarly to studying the vibration movement, the molecule can be replaced by a fictitious particle, with a mass equal to its reduced mass μ, turning around a point at the fixed distance r^e and with angular coordinates φ and θ, respectively, varying from 0 to 2π and from 0 to π.

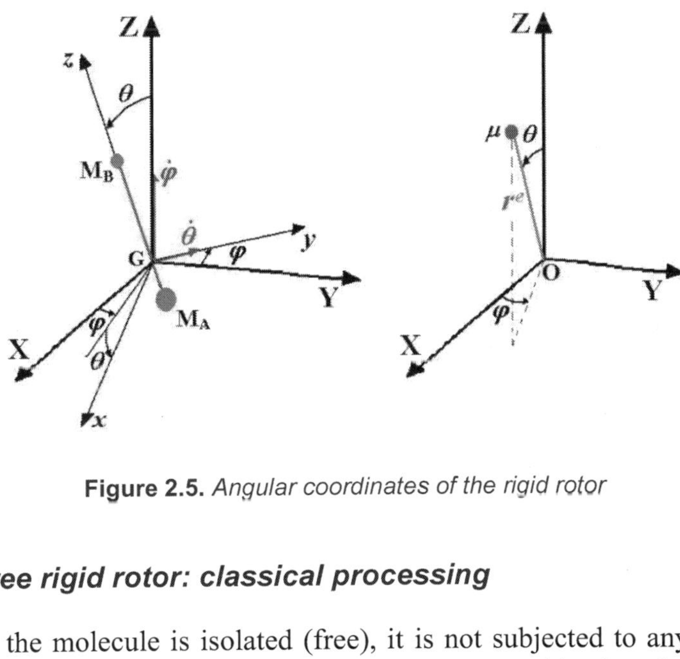

Figure 2.5. *Angular coordinates of the rigid rotor*

2.3.1. *Free rigid rotor: classical processing*

When the molecule is isolated (free), it is not subjected to any potential energy. Its total energy is equal to its kinetic energy and is given by:

$$E_{\text{rot}} = T_{\text{rot}} = \frac{1}{2} I^e (\dot{\theta}^2 + \dot{\phi}^2 \sin^2 \theta) = \frac{p_\theta^2}{2I^e} + \frac{p_\phi^2}{2I^e \sin^2 \theta} \qquad [2.79]$$

where I^e is the molecule's moment of inertia with respect to an axis perpendicular to the internuclear axis and passing through its center of mass:

$$I^e = M_A r_A^{e2} + M_B r_B^{e2} = \mu . r^{e2} \qquad [2.80]$$

$\dot{\phi}$ and $\dot{\theta}$ are the angular velocity components of the molecule's rotation movement, and p_ϕ and p_θ are the conjugate momenta of the angular variables ϕ and θ:

$$p_\theta = \frac{\partial T_{\text{rot}}}{\partial \dot{\theta}} = I^e \dot{\theta}$$
$$p_\phi = \frac{\partial T_{\text{rot}}}{\partial \dot{\phi}} = I^e \dot{\phi} \sin^2 \theta \qquad [2.81]$$

2.3.2. *Free rigid rotor: quantum aspect*

Just as for the vibrational movement, the absorption or emission spectrum of the rigid rotor is composed of well-separated lines. The molecule's rotation movement is quantified. The quantum state of the free rotor is described by the stationary rotational wave function $\Psi_{\text{rot}}(\phi,\theta)$.

The quantification of the rotation movement is obtained by replacing the conjugate momenta [2.81] with the operators $p_\phi = -i\hbar \frac{\partial}{\partial \phi}$ and $p_\theta = -i\hbar \frac{\partial}{\partial \theta}$. The rotational Hamiltonian operator is therefore written as:

$$H_{\text{rot}} = T_{\text{rot}} = \frac{\mathbf{J}^2}{2I^e} = -\frac{\hbar^2}{2I^e}\left(\frac{\partial^2}{\partial \theta^2} + \frac{\cos\theta}{\sin\theta}\frac{\partial}{\partial \theta} + \frac{1}{\sin^2\theta}\frac{\partial^2}{\partial \phi^2}\right) \qquad [2.82]$$

In this equation, **J** is a Hermitian operator called "kinetic momentum" or "angular momentum" linked to the molecule's rotation movement [COH 73, BAR 71]. Its components in the fixed frame are given by:

$$\mathbf{J}_X = -i\hbar\left(-\sin\phi \frac{\partial}{\partial \theta} - \frac{\cos\theta}{\sin\theta}\cos\phi \frac{\partial}{\partial \phi}\right) \qquad [2.83a]$$

$$\mathbf{J}_Y = -i\hbar\left(\cos\phi \frac{\partial}{\partial \theta} - \frac{\cos\theta}{\sin\theta}\sin\phi \frac{\partial}{\partial \phi}\right) \qquad [2.83b]$$

$$\mathbf{J}_Z = -i\hbar\frac{\partial}{\partial\phi} \quad [2.83c]$$

The eigenfunctions and eigenenergies of the rotational Hamiltonian operator are determined by solving the Schrödinger equation. The rotational eigenstates are characterized by two quantum numbers (*J*, *M*).

Thus, the eigenenergy associated with an eigenstate (*J*, *M*) is given by:

$$E_J = \frac{\hbar^2}{2I^e}J(J+1) = B^e J(J+1) \quad [2.84]$$

where $B^e = \hbar^2/2I^e$ is the rotational constant of the rigid molecule. It has the dimension of an energy and, in spectroscopy, it is generally expressed in cm^{-1}.

The corresponding eigenfunction is the product of two functions, one depending on the angle θ and the other on the angle φ :

$$Y_J^M(\theta,\phi) = \Theta_{JM}(\theta)\Phi_M(\phi) \quad [2.85]$$

also written as $|JM\rangle$, in which

$$\Theta_{JM}(\theta) = (-1)^M\sqrt{\frac{(2J+1)}{2}\frac{(J-M)!}{(J+M)!}}P_J^M(\cos\theta) \quad [2.86]$$

with $P_J^M(\cos\theta)$ being the Legendre function defined by:

$$P_J^M(\cos\theta) = \frac{1}{2^J J!}(\sin\theta)^M\frac{d^{J+M}}{d(\cos\theta)^{J+M}}\left(\left(\cos^2\theta - 1\right)^J\right) \quad [2.87]$$

and

$$\Phi_M(\phi) = \frac{1}{\sqrt{2\pi}}e^{iM\phi} \quad [2.88]$$

All of these eigenfunctions make up an orthonormal basis called the "spherical harmonic basis", which are orthonormal and square integrable, and, which verify some symmetry properties [EDM 57, COH 73, BAR 71]. The orthogonality property is given by:

$$\langle J'M'|JM\rangle = \int_0^{2\pi} d\phi \int_0^{\pi} Y_{J'}^{M'*}(\theta,\phi) Y_J^M(\theta,\phi)\sin\theta d\theta = \delta_{MM'}\delta_{JJ'} \quad [2.89]$$

where δ is the Kronecker symbol.

The quantum numbers J and M are such that: J = 0, 1, 2, ... (natural integer) and M = - J, - J + 1,...., - 2, - 1, 0, 1, 2, ..., J - 1, J. This indicates that for a value of the quantum number J, the quantum number M takes (2J + 1) values. It is said that the eigenstate has a degeneracy equal to (2J + 1). In other words, to each state of energy E_J are associated (2J + 1) substates corresponding to (2J + 1) eigenfunctions. This degeneracy can be removed by applying a magnetic field. This is why the quantum number M is called the "magnetic quantum number".

The expressions of the first spherical harmonics are given by:

$$Y_0^0 = \frac{1}{2\sqrt{\pi}} \quad [2.90a]$$

$$Y_1^0 = \frac{1}{2}\sqrt{\frac{3}{\pi}}\cos\theta \quad [2.90b]$$

$$Y_1^{\pm 1} = \mp\frac{1}{2}\sqrt{\frac{3}{2\pi}}\sin\theta e^{\pm i\phi} \quad [2.90c]$$

$$Y_2^0 = \frac{1}{4}\sqrt{\frac{5}{\pi}}(3\cos^2\theta - 1) \quad [2.90d]$$

$$Y_2^{\pm 1} = \mp\frac{1}{2}\sqrt{\frac{15}{2\pi}}\sin\theta\cos\theta e^{\pm i\phi} \quad [2.90e]$$

$$Y_2^{\pm 2} = \frac{1}{4}\sqrt{\frac{15}{2\pi}}\sin^2\theta e^{\pm 2i\phi} \qquad [2.90f]$$

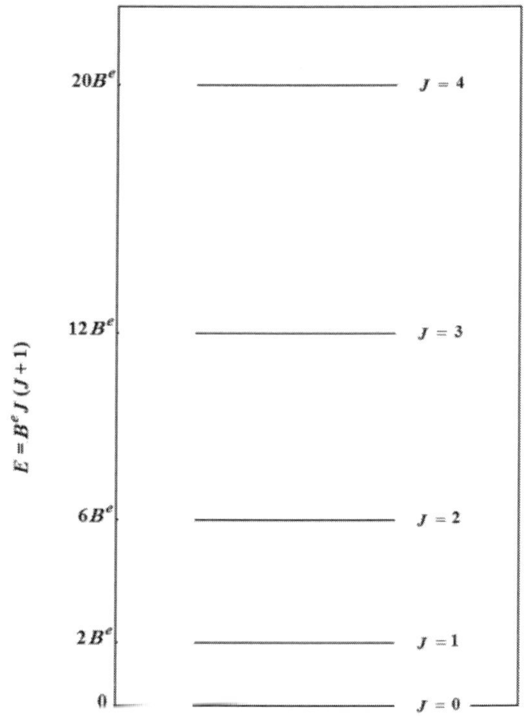

Figure 2.6. *First-energy levels of the rigid rotor*

Figure 2.6 shows the first-energy levels of the rigid rotor. These levels are not equidistant. The energy gap between the two consecutive levels, J and J+1, is:

$$\Delta E_{J,J+1} = B^e\left[(J+1)(J+2) - J(J+1)\right] = 2B^e(J+1) \qquad [2.91]$$

and increases linearly with J.

2.3.3. *Transitions between rotational levels: selection rules*

When the rigid molecule makes a pure rotational transition between one initial state $\left|J_i M_i\right\rangle$ of energy E_{J_i} and a final state $\left|J_f M_f\right\rangle$ of energy E_{J_f}, it absorbs or emits a photon of energy $\Delta E_{J_i,J_f}$ depending on whether $E_{J_i} \langle E_{J_f}$ or $E_{J_i} \rangle E_{J_f}$. This is expressed by the presence of a line in the infrared spectrum in the microwave region. Its dipole moment $\vec{\mu}$ coincides with its permanent dipole moment $\vec{\mu}^e$, but changes direction with respect to the fixed frame (which will be called $\vec{\mu}_A$ in Chapter 4) and thus is split into three components μ_X^e μ_Y^e and μ_Z^e:

$$\vec{\mu}^e = \begin{pmatrix} \mu_X^e \\ \mu_Y^e \\ \mu_Z^e \end{pmatrix} = \mathrm{M}^{-1}(\phi,\theta) \begin{pmatrix} 0 \\ 0 \\ \mu^e \end{pmatrix} = \mu^e \begin{pmatrix} \sin\theta\cos\phi \\ \sin\theta\sin\phi \\ \cos\theta \end{pmatrix}, \qquad [2.92]$$

where μ^e is the value of the rigid molecule's dipole moment along the axis of the bond and $\mathrm{M}^{-1}(\varphi,\theta)$ is the inverse of the matrix $\mathrm{M}(\phi,\theta)$, which transforms a vector from the fixed frame to the mobile frame bound to the molecule (Appendix, section 2.5).

The rotational transition is only possible if the transition moment is *non-zero*, that is:

$$\left\langle J_f M_f \left| \vec{\mu}^e \right| J_i M_i \right\rangle = \int_0^{2\pi} \int_0^{\pi} Y_{J_f}^{M_f *}(\theta,\phi) \vec{\mu}^e Y_{J_i}^{M_i}(\theta,\phi) \sin\theta d\theta d\phi \neq 0 \qquad [2.93]$$

It is clear that this transition moment is always zero in the case of homonuclear molecules such as H_2, N_2, O_2, etc. Indeed, these molecules do not have a permanent dipole moment $\mu^e = 0$ and therefore do not present pure rotational spectra.

In the case of heteronuclear molecules, determining the transition moment leads to calculating the matrix elements of the three components μ_X^e, μ_Y^e and μ_Z^e (see Appendix, section 2.5). The only non-zero terms are:

$$\langle J_i+1, M_i|\cos\theta|J_i M_i\rangle = \sqrt{\frac{(J_i+1)^2 - M_i^2}{(2J_i+1)(2J_i+3)}} \quad [2.94a]$$

leading to the selection rules $\Delta J = \pm 1$ and $\Delta M = 0$;

$$\langle J_i+1, M_i \pm 1|\sin\theta\cos\phi|J_i M_i\rangle = \langle J_i+1, M_i \pm 1|\sin\theta\sin\phi|J_i M_i\rangle$$
$$= \frac{1}{2}\sqrt{\frac{(J_i \pm M_i+1)(J_i \pm M_i+2)}{(2J_i+1)(2J_i+3)}} \quad [2.94b]$$

leading to the selection rules $\Delta J = \pm 1$ and $\Delta M = \pm 1$. However, given that the energy associated with an eigenstate only depends on the quantum number J, only the transitions $\Delta J = \pm 1$ appear on the spectra.

2.4. Vibration–rotation coupling of a free diatomic molecule

2.4.1. *Non-rigid rotor*

In this part, the molecule is again assumed to be in its fundamental electronic state. The translation movements are ignored again. The origin of the fixed or mobile frame is placed at the molecule's center of mass.

If one takes into account the fact that the molecule strains throughout its rotation movement, one is led to simultaneously study the vibration–rotation or rovibrational movements.

We are talking about the non-rigid rotor.

The vibration–rotation Hamiltonian operator of the free molecule is written as:

$$H_{\text{vib-rot}} = -\frac{\hbar^2}{2\mu r^2}\left(\frac{\partial}{\partial r}\left(r^2\frac{\partial}{\partial r}\right) + \frac{\partial^2}{\partial\theta^2} + \frac{\cos\theta}{\sin\theta}\frac{\partial}{\partial\theta} + \frac{1}{\sin^2\theta}\frac{\partial^2}{\partial\phi^2}\right) + V_{\text{vib}} \quad [2.95]$$

where r is the instantaneous internuclear distance (vibrational internal coordinate) and V is the potential energy function of vibration defined in equation [2.13].

However, given the energy difference between vibration and rotation, it is possible to use a perturbative calculation in order to determine which corrections to apply to the previous results.

Solving the Schrödinger equation associated with the rovibrational movement and using the perturbative calculation up to the second order leads to the eigenenergies:

$$E_{\mathrm{v},J} = E_{\mathrm{v}} + B^{\mathrm{v}} J(J+1) - D^{\mathrm{v}} \left[J(J+1)\right]^2 \quad [2.96]$$

where E_{v} is the eigenenergy associated with the anharmonic oscillator given in equation [2.47] and where,

$$B^{\mathrm{v}} = B^e - \alpha^e(\mathrm{v}+\tfrac{1}{2}) \quad [2.97a]$$

$$D^{\mathrm{v}} = D^e + \gamma^e(\mathrm{v}+\tfrac{1}{2}) \quad [2.97b]$$

are, respectively, the molecule's rotational constant and centrifugal distortion constant, depending on the vibrational state in which it is found. α^e and γ^e are the constants characterizing the vibration–rotation coupling [SUG 65]. The corresponding eigenfunctions are:

$$\Psi_{\mathrm{v},JM}(r,\theta,\phi) = \left|\mathrm{v},JM\right\rangle = \left|\mathrm{v}\right\rangle \otimes \left|JM\right\rangle_{\mathrm{v}} \quad [2.98]$$

where $\left|\mathrm{v}\right\rangle$ is the eigenfunction associated with the anharmonic oscillator obtained previously (equation [2.49]) and $\left|JM\right\rangle_{\mathrm{v}}$ is the eigenfunction of the non-rigid rotor depending on the vibrational state in which the molecule is found.

This is represented by a distribution of the rotational states slightly different from the rigid rotor case between the vibrational states.

2.4.2. *Rovibrational transitions: selection rules*

For a rovibrational transition $\left|\mathrm{v}_i, J_i M_i\right\rangle \rightarrow \left|\mathrm{v}_f, J_f M_f\right\rangle$ to take place, the transition moment needs to be *non-zero*. This requires the rotational and vibrational transition moments to be simultaneously *non-zero*. If the series expansion of the molecule's dipole moment is truncated in the first order, we have:

$$\langle \mathrm{v}_f, J_f M_f | \vec{\mu} | \mathrm{v}_i, J_i M_i \rangle = \langle \mathrm{v}_f | q | \mathrm{v}_i \rangle \times {}_{\mathrm{v}_f}\langle J_f M_f | \left(\frac{d\vec{\mu}}{dq} \right)_{q=0} | J_i M_i \rangle_{\mathrm{v}_i} \neq 0 \quad [2.99]$$

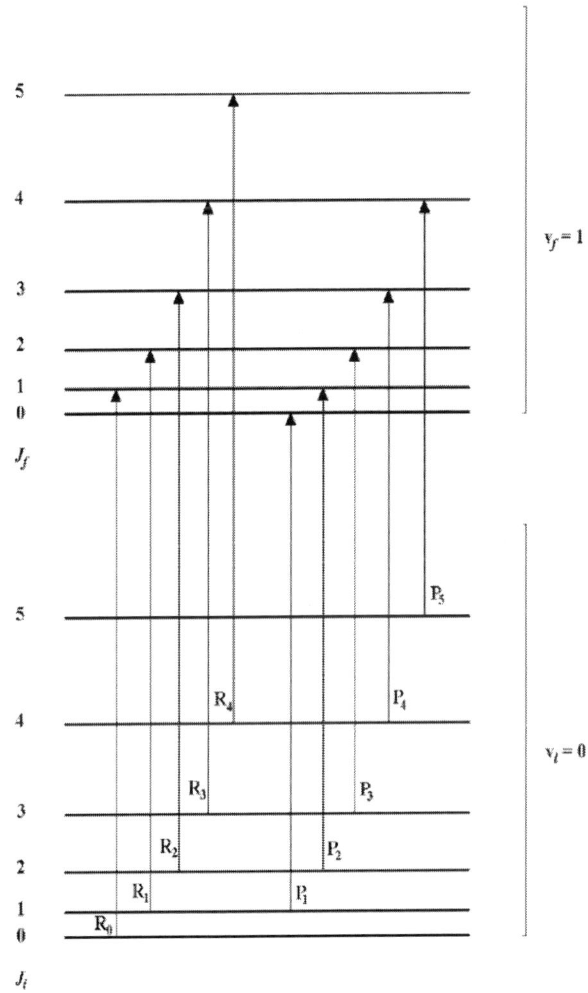

Figure 2.7. *Rovibrational transitions in absorption ($\Delta \mathrm{v} = +1$). The rotational transitions $\Delta J = +1$ and $\Delta J = -1$ are associated with the branches R and P, respectively*

We are then led to express the vector $\left(\dfrac{d\vec{\mu}}{dq}\right)_{q=0}$ in the fixed frame (G, $\vec{X}$, $\vec{Y}$, $\vec{Z}$) in the same way as for the permanent dipole moment $\vec{\mu}^e$ in equation [2.92]. We have:

$$\left(\frac{d\vec{\mu}}{dq}\right)_{q=0} = M^{-1}(\phi,\theta)\begin{pmatrix} 0 \\ 0 \\ \left(\dfrac{d\mu}{dq}\right)_{q=0} \end{pmatrix} = \left(\frac{d\mu}{dq}\right)_{q=0}\begin{pmatrix} \sin\theta\cos\phi \\ \sin\theta\sin\phi \\ \cos\theta \end{pmatrix} \qquad [2.100]$$

Equations [2.99] and [2.100] show that the rovibrational selection rules are:

$$\Delta v = \pm 1 \qquad \text{and} \qquad \Delta J = \pm 1$$

with the transition $\Delta v = 0$ being forbidden. In the rovibrational spectra, the lines of the branch R (respectively P) are on the high (respectively low) frequencies side with respect to the central vibrational frequency.

2.5. Appendix

2.5.1. *The commutators*

$$[N,a] = \left[a^{\dagger}a,a\right] = \left[a^{\dagger},a\right]a = -a$$

$$\left[N,a^{\dagger}\right] = \left[a^{\dagger}a,a^{\dagger}\right] = a^{\dagger}\left[a,a^{\dagger}\right] = a^{\dagger}$$

2.5.2. *Expressions of p^n and q^n in terms of the operators a and $a^{\dagger}$*

$$q^2 = \frac{1}{2}\left(a^{\dagger 2} + a^2 + (2N+1)\right)$$

$$q^3 = \frac{1}{2\sqrt{2}}\left(a^{\dagger 3} + a^3 + 3Na^{\dagger} + 3(N+1)a\right)$$

$$q^4 = \frac{1}{4}\left(a^{\dagger 4} + a^4 + 2(2N-1)a^{\dagger 2} + 2(2N+3)a^2 + 3(2N^2+2N+1)\right)$$

$$q^5 = \frac{1}{4\sqrt{2}}\left(a^{\dagger 5} + a^5 + 5(N-1)a^{\dagger 3} + 5(N+2)a^3 + 5(2N^2+1)a^{\dagger} + 5(2N^2+4N+3)a\right)$$

$$p^2 = -\frac{1}{2}\left(a^{\dagger 2} + a^2 - (2N+1)\right)$$

$$p^3 = -\frac{i}{2\sqrt{2}}\left(a^{\dagger 3} - a^3 - 3Na^{\dagger} + 3(N+1)a\right)$$

$$p^4 = \frac{1}{4}\left(a^{\dagger 4} + a^4 - 2(2N-1)a^{\dagger 2} - 2(2N+3)a^2 + 3(2N^2+2N+1)\right)$$

$$p^5 = \frac{i}{4\sqrt{2}}\left(a^{\dagger 5} - a^5 - 5(N-1)a^{\dagger 3} + 5(N+2)a^3 + 5(2N^2+1)a^{\dagger} - 5(2N^2+4N+3)a\right)$$

2.5.3. *Matrix elements of p^n and q^n*

The matrix elements of p^n and q^n (n = 1, 2, 3, 4) can be found on the basis of their expressions in terms of the operators a and $a^{\dagger}$:

$$\langle v|q|v-1\rangle = \sqrt{\frac{v}{2}} \qquad \langle v|p|v-1\rangle = i\sqrt{\frac{v}{2}}$$

$$\langle v|q|v+1\rangle = \sqrt{\frac{v+1}{2}} \qquad \langle v|p|v+1\rangle = -i\sqrt{\frac{v+1}{2}}$$

$$\langle v|q^2|v-2\rangle = \frac{1}{2}\sqrt{v(v-1)} \qquad \langle v|p^2|v-2\rangle = -\frac{1}{2}\sqrt{v(v-1)}$$

$$\langle v|q^2|v\rangle = v + \frac{1}{2} \qquad \langle v|p^2|v\rangle = v + \frac{1}{2}$$

$$\langle \mathrm{v}|q^2|\mathrm{v}+2\rangle = \frac{1}{2}\sqrt{(\mathrm{v}+1)(\mathrm{v}+2)}$$

$$\langle \mathrm{v}|p^2|\mathrm{v}+2\rangle = -\frac{1}{2}\sqrt{(\mathrm{v}+1)(\mathrm{v}+2)}$$

$$\langle \mathrm{v}|q^3|\mathrm{v}-3\rangle = \left(\frac{\mathrm{v}(\mathrm{v}-1)(\mathrm{v}-2)}{8}\right)^{1/2}$$

$$\langle \mathrm{v}|p^3|\mathrm{v}-3\rangle = -i\left(\frac{\mathrm{v}(\mathrm{v}-1)(\mathrm{v}-2)}{8}\right)^{1/2}$$

$$\langle \mathrm{v}|q^3|\mathrm{v}-1\rangle = 3\left(\frac{\mathrm{v}}{2}\right)^{3/2} \qquad \langle \mathrm{v}|p^3|\mathrm{v}-1\rangle = 3i\left(\frac{\mathrm{v}}{2}\right)^{3/2}$$

$$\langle \mathrm{v}|q^3|\mathrm{v}+1\rangle = 3\left(\frac{\mathrm{v}+1}{2}\right)^{3/2} \qquad \langle \mathrm{v}|p^3|\mathrm{v}+1\rangle = -3i\left(\frac{\mathrm{v}+1}{2}\right)^{3/2}$$

$$\langle \mathrm{v}|q^3|\mathrm{v}+3\rangle = \left(\frac{(\mathrm{v}+1)(\mathrm{v}+2)(\mathrm{v}+3)}{8}\right)^{1/2}$$

$$\langle \mathrm{v}|p^3|\mathrm{v}+3\rangle = i\left(\frac{(\mathrm{v}+1)(\mathrm{v}+2)(\mathrm{v}+3)}{8}\right)^{1/2}$$

$$\langle \mathrm{v}|q^4|\mathrm{v}-4\rangle = \frac{1}{4}\left(\mathrm{v}(\mathrm{v}-1)(\mathrm{v}-2)(\mathrm{v}-3)\right)^{1/2}$$

$$\langle \mathrm{v}|p^4|\mathrm{v}-4\rangle = \frac{1}{4}\left(\mathrm{v}(\mathrm{v}-1)(\mathrm{v}-2)(\mathrm{v}-3)\right)^{1/2}$$

$$\langle \mathrm{v}|q^4|\mathrm{v}-2\rangle = \frac{1}{2}(2\mathrm{v}-1)\left(\mathrm{v}(\mathrm{v}-1)\right)^{1/2}$$

$$\langle \mathrm{v}|p^4|\mathrm{v}-2\rangle = -\frac{1}{2}(2\mathrm{v}-1)\left(\mathrm{v}(\mathrm{v}-1)\right)^{1/2}$$

$$\langle \mathrm{v}|q^4|\mathrm{v}\rangle = \frac{3}{2}\left(\mathrm{v}+\frac{1}{2}\right)^2+\frac{3}{8} \qquad \langle \mathrm{v}|p^4|\mathrm{v}\rangle = \frac{3}{2}\left(\mathrm{v}+\frac{1}{2}\right)^2+\frac{3}{8}$$

$$\langle \mathrm{v}|q^4|\mathrm{v}+2\rangle = \frac{1}{2}(2\mathrm{v}+3)\left((\mathrm{v}+1)(\mathrm{v}+2)\right)^{1/2}$$

$$\langle \mathrm{v}|p^4|\mathrm{v}+2\rangle = -\frac{1}{2}(2\mathrm{v}+3)\left((\mathrm{v}+1)(\mathrm{v}+2)\right)^{1/2}$$

$$\langle \mathrm{v}|q^4|\mathrm{v}+4\rangle = \frac{1}{4}\left((\mathrm{v}+1)(\mathrm{v}+2)(\mathrm{v}+3)(\mathrm{v}+4)\right)^{1/2}$$

$$\langle \mathrm{v}|p^4|\mathrm{v}+4\rangle = \frac{1}{4}\left((\mathrm{v}+1)(\mathrm{v}+2)(\mathrm{v}+3)(\mathrm{v}+4)\right)^{1/2}$$

$$\langle \mathrm{v}|pq|\mathrm{v}-2\rangle = \frac{i}{2}\sqrt{(\mathrm{v}-1)(\mathrm{v})} \quad \langle \mathrm{v}|qp|\mathrm{v}-2\rangle = \frac{i}{2}\sqrt{(\mathrm{v}-1)(\mathrm{v})}$$

$$\langle \mathrm{v}|pq|\mathrm{v}\rangle = -\frac{\mathrm{i}}{2} \qquad \langle \mathrm{v}|qp|\mathrm{v}\rangle = \frac{\mathrm{i}}{2}$$

$$\langle \mathrm{v}|pq|\mathrm{v}+2\rangle = -\frac{i}{2}\sqrt{(\mathrm{v}+1)(\mathrm{v}+2)}$$

$$\langle \mathrm{v}|qp|\mathrm{v}+2\rangle = -\frac{i}{2}\sqrt{(\mathrm{v}+1)(\mathrm{v}+2)}$$

$$\langle \mathrm{v}|p^2q^2|\mathrm{v}-2\rangle = \sqrt{(\mathrm{v}-1)\mathrm{v}} \qquad \langle \mathrm{v}|q^2p^2|\mathrm{v}-2\rangle = -\sqrt{(\mathrm{v}-1)\mathrm{v}}$$

$$\langle \mathrm{v}|p^2q^2|\mathrm{v}\rangle = \frac{1}{2}\left(\mathrm{v}+\frac{1}{2}\right)^2-\frac{3}{8} \qquad \langle \mathrm{v}|q^2p^2|\mathrm{v}\rangle = \frac{1}{2}\left(\mathrm{v}+\frac{1}{2}\right)^2-\frac{3}{8}$$

$$\langle \mathrm{v}|p^2q^2|\mathrm{v}+2\rangle = -\sqrt{(\mathrm{v}+1)(\mathrm{v}+2)} \qquad \langle \mathrm{v}|q^2p^2|\mathrm{v}+2\rangle = \sqrt{(\mathrm{v}+1)(\mathrm{v}+2)}$$

$$\langle \mathrm{v}|pq^3|\mathrm{v}-2\rangle = \frac{i}{2}\left(\mathrm{v}(\mathrm{v}-1)(\mathrm{v}-2)\right)^{1/2}$$

$$\langle \mathrm{v}|q^3p|\mathrm{v}-2\rangle = \frac{i}{2}(\mathrm{v}(\mathrm{v}-1)(\mathrm{v}+1))^{1/2}$$

$$\langle \mathrm{v}|pq^3|\mathrm{v}\rangle = -\frac{3\mathrm{i}}{2}\left(\mathrm{v}+\frac{1}{2}\right) \qquad \langle \mathrm{v}|q^3p|\mathrm{v}\rangle = \frac{3\mathrm{i}}{2}\left(\mathrm{v}+\frac{1}{2}\right)$$

$$\langle \mathrm{v}|pq^3|\mathrm{v}+2\rangle = -\frac{i}{2}((\mathrm{v}+1)(\mathrm{v}+2)(\mathrm{v}+3))^{1/2}$$

$$\langle \mathrm{v}|q^3p|\mathrm{v}+2\rangle = -\frac{i}{2}(\mathrm{v}(\mathrm{v}+1)(\mathrm{v}+2))^{1/2}$$

$$\langle \mathrm{v}|p^3q|\mathrm{v}-2\rangle = \frac{i}{2}(\mathrm{v}(\mathrm{v}-1)(\mathrm{v}+1))^{1/2}$$

$$\langle \mathrm{v}|qp^3|\mathrm{v}-2\rangle = \frac{i}{2}(\mathrm{v}(\mathrm{v}-1)(\mathrm{v}-2))^{1/2}$$

$$\langle \mathrm{v}|p^3q|\mathrm{v}\rangle = -\frac{3\mathrm{i}}{2}\left(\mathrm{v}+\frac{1}{2}\right)$$

$$\langle \mathrm{v}|qp^3|\mathrm{v}\rangle = \frac{3\mathrm{i}}{2}\left(\mathrm{v}+\frac{1}{2}\right)$$

$$\langle \mathrm{v}|p^3q|\mathrm{v}+2\rangle = -\frac{i}{2}(\mathrm{v}(\mathrm{v}+1)(\mathrm{v}+2))^{1/2}$$

$$\langle \mathrm{v}|qp^3|\mathrm{v}+2\rangle = -\frac{i}{2}((\mathrm{v}+1)(\mathrm{v}+2)(\mathrm{v}+3))^{1/2}$$

2.5.4. *Matrix of rotation and rotational transitions*

The unitary rotational matrix transformation from the fixed frame (G, $\vec{X}$, $\vec{Y}$, $\vec{Z}$) to the mobile frame bound to the diatomic (linear) molecule (G, $\vec{x}$, $\vec{y}$, $\vec{z}$) is expressed as:

$$\mathrm{M}(\phi,\theta) = \begin{pmatrix} \cos\phi\cos\theta & \sin\phi\cos\theta & -\sin\theta \\ -\sin\phi & \cos\phi & 0 \\ \cos\phi\sin\theta & \sin\phi\sin\theta & \cos\theta \end{pmatrix}$$

The rotational transition moments of the dipole moment components are given by:

$$\langle J_f M_f | \sin\theta\cos\phi | J_i M_i \rangle = \int_0^{2\pi}\int_0^{\pi} Y_{J_f}^{M_f *}(\theta,\phi)\sin\theta\cos\phi Y_{J_i}^{M_i}(\theta,\phi)\sin\theta d\theta d\phi$$

$$\langle J_f M_f | \sin\theta\sin\phi | J_i M_i \rangle = \int_0^{2\pi}\int_0^{\pi} Y_{J_f}^{M_f *}(\theta,\phi)\sin\theta\sin\phi Y_{J_i}^{M_i}(\theta,\phi)\sin\theta d\theta d\phi$$

$$\langle J_f M_f | \cos\theta | J_i M_i \rangle = \int_0^{2\pi}\int_0^{\pi} Y_{J_f}^{M_f *}(\theta,\phi)\cos\theta Y_{J_i}^{M_i}(\theta,\phi)\sin\theta d\theta d\phi$$

To determine these elements, it is required to, on the one hand, express the terms $\sin\theta\cos\phi$, $\sin\theta\sin\phi$ and $\cos\theta$ in terms of the spherical harmonics Y_1^0, Y_1^{+1} and Y_1^{-1}. This exercise is left to the reader. On the other hand, the composition rule of the spherical harmonics that allows these moments to be calculated is expressed as:

$$Y_{J_1}^{M_1}(\theta,\phi)Y_{J_2}^{M_2}(\theta,\phi) = \sum_{J=|J_1-J_2|}^{J_1+J_2} \sqrt{\frac{(2J_1+1)(2J_2+1)}{4\pi(2J+1)}} C(J_1 J_2 J; M_1 M_2) \times C(J_1 J_2 J; 00) Y_J^{M_1+M_2}(\theta,\phi)$$

The coefficients C are called "Clebsch–Gordan coefficients". Their expression can be found in [ROS 67].

3

Profile and Shape of Spectral Lines

The characteristics of an absorption or emission line of a diatomic molecule depend on the mechanisms that compete to broaden or narrow its width, of which the main ones are the velocity, which depends on the temperature, and the soft or hard collisions experienced under the effect of the pressure. The natural width of a vibration–rotation energy level, connected to its lifetime, is of Lorentzian type. The Doppler–Fizeau type broadening observed in a fixed reference frame for the active molecule is connected to its relative velocity. Collisions reduce the lifetime of the energy levels and in the case of isolated lines lead to broadening. When the two mechanisms, velocity and collision effects, are not negligible, the line profile, a convolution product of the Doppler and the Lorentz profiles, is of Voigt type. The narrowing phenomena are due to the Dicke–Galatry effect that leads to the Galatry–Nelkin–Ghatak–Rautian–Sobelmann profiles or, in the case of non-isolated lines, to interference phenomena. In frequency-resolved spectroscopy (absorption, emission, Rayleigh diffusion, Raman diffusion, etc.), the active molecule is subjected to an electromagnetic radiation, while it is in interaction with other molecules that constitute a thermal bath in the sense of statistical physics. To interpret the width and the shift of the isolated lines, the Robert and Bonamy (RB) theory described in this chapter expresses the macroscopic effect of this bath on the spectral lines in terms of a density operator and binary collision parameters, which are assumed to be sudden (impact approximation) according to rectilinear or parabolic classical trajectories (semiclassical theories). Its expansion in the Liouville lines space makes it possible to apply the linked cluster theorem to calculate the statistical effects of the thermal bath through the cumulants expansion by including the usual electrostatic forces, the short- and mid-range interactions (atom-atom potential), to describe the collision effects. The width and shift of a line are calculated using resonance functions based on the collisional differential cross-section determined by the RB theory.

3.1. Introduction

The study of line broadening by pressure is expressed relative to the distribution of the radiation emitted or absorbed by molecules, which undergo collisions with other similar or different molecules.

Many semiclassical theories have been proposed calculating the broadening parameter of a line due to collisions. In these theories, the molecule's vibration and rotation degrees of freedom are processed from a quantum point of view, whereas their translation movement is envisaged from a classical point of view.

The ATC model, built on the Anderson theory [AND 49] and generalized by Tsao and Curnutte [TSA 62], has been widely used to calculate the broadening coefficient of a spectral line by pressure. We recall that in the binary collision model proposed by these authors, the relative trajectory of two molecules partners of a collision is assumed to be rectilinear and the differential cross-section is calculated using the second-order perturbation theory. This model is easy to apply, which explains why it has been widely used. However, it does have the disadvantage of being unrealistic when the collisions are effective in the short range and to lead to a calculated value of the differential cross-section, which diverges for collisions characterized by an impact parameter tending toward zero. This difficulty can be cancelled out by a cut-off method, by considering any collision characterized by an impact parameter lower than a critical value as being totally effective.

Thereafter, we are going to recall the principle of the calculations of the broadening parameter essentially based on the Robert and Bonamy [ROB 79] semiclassical theoretical model following the works of Baranger, Kolb, Griem, Fano and Ben Reuven [BAR 58, BEN 75, FAN 63, KOL 58]. To calculate the spectral distribution of the absorbed radiation, they used the Liouville formalism.

In this chapter, we introduce the different steps of the theoretical development to determine the expression of the differential cross-section, which we are led to define in a semiclassical collision theory. Within the framework of the impact approximation, meaning that the collisions are binary and instantaneous, the differential cross-section is expressed according to the diffusion operator of the physical system considered (an active molecule and a perturbing molecule, partners of a collision). By using the linked cluster theorem to calculate the cumulant expansion of the diffusion operator up to second order in the "Liouville space" (double states space or lines space), Robert and Bonamy obtain an expression of the differential cross-section, which avoids the cut-off procedure used in the Andersson–Tsao–Curnutte theory.

3.2. Semiclassical model of calculating the broadening parameters of spectral lines

These theories generally lead to an expression of the broadening parameter in terms of a differential cross-section (outlined in section 3.2.5), which is first calculated for a given collision, and then averaged on all the geometries of all possible collisions. The term "semiclassical" refers to the classical processing of the translation movement.

3.2.1. *General description of the interacting physical system*

The system considered is a gas composed of one or several molecular species, enclosed in a volume V and immersed in a radiation field. As it is prohibitive to calculate the eigenstates of this system, it is assumed that the optically active molecules are diluted in an inactive perturbing bath, constituting a thermostat with which they interact.

In such a situation, the physical system can be considered as a set of three subsystems whose states' spaces are composed of the eigenvectors of the non-perturbed Hamiltonians, respectively, H_A of the active molecules, H_B of the molecules of the thermal bath and H_R of the electromagnetic radiation.

Considering the interactions between these subsystems, and because the molecules of the bath do not interact with the electromagnetic radiation, the physical system's Hamiltonian can be written as:

$$H = H_A + H_B + H_{AB} + H_R + H_{AR} \quad [3.1]$$

where H_{AB} and H_{AR} are the interaction Hamiltonians, respectively, between the active molecules and the molecules of the bath, and between the active molecules and the radiation.

In this general form, the expression of the total Hamiltonian H is still too complicated to be resolved. Some hypotheses and approximations are necessary to simplify the processing, in particular to consider only the binary interactions, restricted to an active molecule and a perturbing molecule, partners in a collision.

3.2.2. *General expression of the profile of a spectral line*

When the matter–radiation coupling is weak (weak fields approximation), the spectral distribution of the radiation absorbed by optically active molecules per volume unit, at the temperature T, is expressed as the Fourier transformation of the time-dependent electrical dipole moment correlation function [BEN 66a, BEN 66b, GOR 68] of all the molecules, also called power spectral density.

The absorption coefficient is expressed as:

$$K(\omega)=\frac{4\pi^2}{3\hbar cV}\omega\left(1-e^{-\hbar\omega/k_BT}\right)I(\omega) \qquad [3.2]$$

with

$$I(\omega)=\lim_{\epsilon\to 0}\frac{1}{2\pi}\int_{-\infty}^{+\infty}e^{-|\epsilon|t}e^{-i\omega t}\left\langle M_z(0)M_z(t)\right\rangle dt \qquad [3.3]$$

also called power spectral density.

In these equations, ω designates the angular frequency of the incident radiation, c is the speed of light, k_B is the Boltzmann constant and $M_z(t)$ is the projection of the electrical dipole moment of all the active molecules following the OZ axis carrying the polarization vector of the electromagnetic field. The symbol $\langle .....\rangle$ represents the quantum average of the product $M_z(0)M_z(t)$ on all the initial conditions at the instant $t = 0$ and all the possible evolutions of the system (composed of active molecules and perturbing molecules of the thermal bath) between the instants 0 and t.

By assuming that the induction phenomena and the correlations between the active molecules are negligible, which is valid for the weak densities of active molecules, this quantum average becomes:

$$\langle M_z(0)M_z(t)\rangle = N\langle \mu_z(0)\mu_z(t)\rangle_0 = N\mathrm{Tr}\left[\rho(0)\mu_z(0)\mu_z(t)\right] \quad [3.4]$$

where N is the number of active molecules and μ_z is the component of an active molecule's electrical dipole parallel to the OZ axis. The trace operator (Tr) now affects the material system composed of N subsystems each including an active molecule and the perturbing molecules of the thermal bath. Under these conditions, the absorption coefficient and the power spectral density are written as:

$$K(\omega) = \frac{4\pi^2 \tilde{N}}{3\hbar c}\omega\left(1 - e^{\hbar\omega/k_B T}\right) I(\omega) \quad [3.5]$$

and

$$I(\omega) = \lim_{\epsilon\to 0}\frac{1}{2\pi}\int_{-\infty}^{+\infty} e^{-|\epsilon| t} e^{-i\omega t}\langle \mu_z(0)\mu_z(t)\rangle dt \quad [3.6]$$

where $\tilde{N} = \frac{N}{V}$ is now the number of optically active molecules per unit volume and $\rho(0)$ of equation [3.4] is the canonical density matrix operator associated with the "active molecule-thermal bath" subsystem. Yet, to denote an active molecule, we will use 1 instead of A.

The Hamiltonian of the new system comprises only three terms:

$$H = H_1 + H_B + V_{1B} \quad [3.7]$$

which, respectively, represent the Hamiltonians of the isolated active molecule, the thermal bath in the absence of the active molecule and the "active molecule–thermal bath" interaction.

In order to facilitate the calculations for solving the Schrödinger equation associated with the Hamiltonian H, Robert and Bonamy used the Liouville formalism (section 3.6.1) [FAN 57, KUB 57] in their model. In this formalism, the Liouville space vectors are defined as tensor products of the

Hilbert space vectors by the conjugate transposed vectors of this same space. In this space, the expression of the dipole moment is given by:

$$\mu_z(t) = e^{i\hbar^{-1}Ht}\mu_z(0)e^{-i\hbar^{-1}Ht} = e^{i\hbar^{-1}Lt}\mu_z(0) \quad [3.8]$$

Just as for the Hamiltonian H in the Hilbert space, the Liouvillian L can be written as:

$$L = L_1 + L_B + L_{1B} \quad [3.9]$$

where in equation [3.10] L_1, L_B and L_{IB} are the Liouvillians associated with the active molecule, the thermal bath and the "active molecule–thermal bath" interaction.

The expression [3.6] can then be written in the following form:

$$I(\omega) = \lim_{\in \to 0} \frac{1}{2\pi} \mathrm{Tr} \int_{-\infty}^{+\infty} e^{-|\in|t} e^{-i\omega t} \left(\mu_z(0) e^{-ih^{-1}Lt} \mu_z(0)\rho(0)\right) dt \quad [3.10]$$

where $\rho(0)$ represents the subsystem in thermodynamic equilibrium in its initial state, and in which we will later put $|\in| = 0$.

The resolvent method enables the calculation of the integral of equation [3.10], by taking the Fourier transform of $e^{-ih^{-1}Lt}$ [FAN 63]. It results that:

$$I(\omega) = -\frac{1}{\pi} \Im m \mathrm{Tr}\left[\mu_z \frac{1}{\omega - L}(\mu_z \rho(0))\right] \quad [3.11]$$

where the trace operation affects all the states of the absorbing molecule and thermal bath subsystem, and $\frac{1}{\omega - L}$ is the resolvent of the operator L that describes the system's dynamics.

From equation [3.9] that can be decomposed in the form $L = L_0 + L_{1B}$ and by using the following relation between operators A and B:

$$\frac{1}{A} - \frac{1}{B} = \frac{1}{B}(B - A)\frac{1}{A} \quad [3.12]$$

the resolvent can be expressed in the form:

$$\frac{1}{\omega-L}=\frac{1}{\omega-L_0}\left[1+M(\omega)\frac{1}{\omega-L_0}\right] \qquad [3.13]$$

where $M(\omega)$ represents the influence of the bath on the active molecule and is expressed as:

$$M(\omega)=L_{1B}\sum_{n=0}^{\infty}\left(\frac{1}{\omega-L_0}L_{1B}\right)^n \qquad [3.14]$$

Furthermore, by assuming that the initial correlations between the active molecule and the thermal bath are negligible (initial chaos hypothesis, the Hamiltonians of the active molecule H_1 and of the thermal bath H_B are statistically independent), one can put the density operator $\rho(0)$ in the form of a product of two density operators $\rho_1(0)\ \rho_B(0)$, such that:

$$\rho(0)=\frac{e-\dfrac{H_1+H_B+V_{1B}}{k_BT}}{\mathrm{Tr}\left[e-\dfrac{H_1+H_B+V_{1B}}{k_BT}\right]}=\rho_1(0)\rho_B(0) \qquad [3.15]$$

with $\rho_1(0)=\dfrac{e-\dfrac{H_1}{k_BT}}{\mathrm{Tr}\left[e-\dfrac{H_1}{k_BT}\right]}$ and $\rho_B(0)=\dfrac{e-\dfrac{H_B}{k_BT}}{\mathrm{Tr}\left[e-\dfrac{H_B}{k_BT}\right]}$

The trace on the material subsystem can be separated into two partial traces:

$$I(\omega)=-\frac{1}{\pi}\Im m\mathrm{Tr}_1\left[\mu_z(0)\frac{1}{\omega-L_1}\left[1+\langle M(\omega)\rangle_B\frac{1}{\omega-L_1}\right](\mu_z(0)\rho_1(0))\right] \qquad [3.16]$$

with $\langle M(\omega)\rangle_B=\mathrm{Tr}_B\left(M(\omega)\rho_B(0)\right)$. Following the thermal equilibrium hypothesis, one has $\rho_B(0)_{ij}=\rho_B(0)\delta_{ij}$ and consequently the contribution of

L_B to L_0 disappears in the operation, which consists of taking the trace on the bath's states.

The introduction of the Zwanzig relaxation operator [ZWA 60] such as:

$$\left\langle M_c(\omega)\right\rangle_B = \left\langle M(\omega)\right\rangle_B \sum_{n=0}^{\infty}\left[\frac{1}{\omega - L_1}\left\langle M(\omega)\right\rangle_B\right]^n \qquad [3.17]$$

leads to the following form for the spectral profile:

$$I(\omega) = -\frac{1}{\pi}\Im m\mathrm{Tr}_1\left[\mu_z(0)\frac{1}{\omega - L_1 - \left\langle M_c(\omega)\right\rangle_B}\left(\mu_z(0)\rho_1(0)\right)\right] \qquad [3.18]$$

The Zwanzig relaxation operator $\left\langle M_c(\omega)\right\rangle_B$ represents the influence of the interaction between the absorbent molecule and the molecules of the bath on the spectral profile and adds as a perturbation to the angular frequency matrix of the radiative transitions of the isolated active molecule.

By only retaining the first term in the expansion of the Zwanzig relaxation operator in the density powers ($n_2 = N_B/V$) of the perturbing gas, it becomes ($\left\langle M_c(\omega)\right\rangle_B \approx n_2\left\langle m(\omega)\right\rangle_B$):

$$I(\omega) = -\frac{1}{\pi}\Im m\mathrm{Tr}_M\left[\mu_z(0)\frac{1}{\omega - L_1 - n_2\left\langle m(\omega)\right\rangle_B}\left(\mu_z(0)\rho_M(0)\right)\right] \qquad [3.19]$$

where $m(\omega)$, which accounts for the interaction between an active molecule and a molecule of the bath, is the solution of the Lippmann–Schwinger equation [LIP 50] (we are brought back to a two-body collision problem, and A and B are, however, noted as 1 and 2, respectively, in the interaction potential of equation [3.7]):

$$m(\omega) = L_{12} + L_{12}\frac{1}{\omega - L_0}L_{12} = L_{12} + L_{12}\frac{1}{\omega - L_0}m(\omega) \qquad [3.20]$$

It expresses the influence of the binary collisions on the spectral profile.

The symbol $\langle\ \rangle_B$ designates an average in the Liouville space on the states of the thermal bath, taking the translation states into account, such that:

$$\langle m(\omega)\rangle_B = \sum_{22'} \langle\langle 2'2' | m(\omega) | 22\rangle\rangle \langle 2 | \rho_2(0) | 2\rangle \qquad [3.21]$$

where $|22\rangle\rangle$ designates a Liouville space vector, which is the tensor product of the Hilbert space by its dual (section 3.6). Finally, by expanding the trace of equation [3.19], the spectral profile becomes:

$$I(\omega) = -\frac{1}{\pi}\sum_{\mathrm{if}}\sum_{\mathrm{i'f'}} \langle \mathrm{i'} | \mu_z(0) | \mathrm{f'}\rangle^* \langle\langle \mathrm{i'f'} \left| \left[\frac{1}{\omega - L_1 - n_2 \langle m(\omega)\rangle_B} \right] \right| \mathrm{if}\rangle\rangle \langle \mathrm{i} | \rho_1(0) | \mathrm{i}\rangle \langle \mathrm{i} | \mu_z(0) | \mathrm{f}\rangle \qquad [3.22]$$

where $|\alpha\rangle = |\mathrm{v}_\alpha \mathrm{J}_\alpha \mathrm{m}_\alpha\rangle$ and $E_\alpha = E_{\mathrm{v}_\alpha \mathrm{J}_\alpha}$ (α = i, f) designate the eigenvectors and the eigenvalues of H_1. In this expression, the operation that consists of taking the trace is performed on the vibration–rotation states corresponding to the active molecule's internal and rotational degrees of freedom. Its translational degrees of freedom are incorporated into those of the thermal bath. This separation is valid because the operators $\mu_z(0)$ and $\rho_1(0)$ are independent of the variables, which characterize the translation of the active system that, in the semiclassical approximation, is processed by classical mechanics.

3.2.3. *Consequences of the invariance of the Zwanzig relaxation operator under rotation*

When the matter–radiation coupling is weak (weak fields approximation), it is legitimate to assume that the active molecule's environment remains isotropic. Under these conditions, the operators $\langle m(\omega)\rangle_B$ and

$\frac{1}{\omega - L_1 - n_2 \langle m(\omega) \rangle_B}$ are invariant under rotation and are therefore scalar operators. To simply express the matrix elements of these operators, a standard basis must be chosen, built on the composition of the kinetic momenta (angular momenta) of the active molecule, partner of a collision such that $J_i - J_f = K_1$, where J_i and J_f designate the rotational angular momenta of the active molecule in the initial state i and in the final state f, respectively. The vectors $\left| v_f J_f m_f v_i J_i m_i \right\rangle\rangle$ and $\left| v_f J_f v_i J_i K_1 M_1 \right\rangle\rangle$ corresponding to the states implied in a transition are connected by the relations, which involve the Clebsh–Gordan coefficients.

It is shown that the matrix elements of a scalar operator T are expressed in the basis $\left| v_f J_f v_i J_i K_1 M_1 \right\rangle\rangle$ in the form:

$$\begin{aligned}\langle\langle v_f^{'} J_f^{'} v_i^{'} J_i^{'} K_1^{'} M_1^{'} \left| T \right| v_f J_f v_i J_i K_1 M_1 \rangle\rangle = \delta_{M_1 M_1^{'}} \delta_{K_1 K_1^{'}} \\ \langle\langle v_f^{'} J_f^{'} v_i^{'} J_i^{'} K_1^{'} M_1^{'} \left| T \right| v_f J_f v_i J_i K_1 M_1 \rangle\rangle\end{aligned} \quad [3.23]$$

The dipole moment operator, which induces the transitions leading to an absorption or an emission, is a vector operator, which is an irreducible tensor operator X_0^1 of rank 1 (X^J with J = 1). For the isotropic Raman diffusion and the anisotropic Raman diffusion, the operator is a tensor of rank 0 with J = 0 for the elastic diffusion (Rayleigh), of rank 2 with J = 2 for the inelastic diffusion (Raman), and of rank 3 with J = 3 for the hyper-Raman processes. In this case, one can only deduce that K_1 = J. The matrix elements of an irreducible tensor of the rank J are calculated using the Wigner–Eckart theorem so that:

$$\langle v'J'm' | X_M^J | v''J''m'' \rangle = C(J''JJ'; m''Mm') \langle v'J' \| X^J \| v''J'' \rangle \quad [3.24]$$

in which the reduced matrix element is expressed as:

$$\langle v'J' \| X^J \| v''J'' \rangle = \sum_{\mu''\mu'} C(J''JJ'; \mu''M'\mu') \langle v'J'\mu' | X_{M'}^J | v''J''\mu'' \rangle \quad [3.25]$$

The orthonormalization properties of the Clebsch–Gordan coefficients enable the sums to be performed on the projection quantum numbers and therefore express the spectral profile of equation [3.22] as follows:

$$I(\omega) = -\frac{1}{\pi}\Im m \sum_{\substack{\mathrm{v_i J_i m_i v_f J_f m_f} \\ \mathrm{v_i' J_i' v_f' J_f'}}} \langle \mathrm{v_i J_i m_i} | \rho_1(0) | \mathrm{v_i J_i m_i} \rangle (-1)^{\mathrm{J_i + J_f'}} \frac{\sqrt{[\mathrm{J_f}][\mathrm{J_f'}]}}{[\mathrm{J}]} \times$$
$$\langle \mathrm{v_f J_f} \| \mathrm{X^J} \| \mathrm{v_i J_i} \rangle \langle \mathrm{v_f' J_f'} \| \mathrm{X^J} \| \mathrm{v_i' J_i'} \rangle^* \langle\langle \mathrm{v_f' J_f' v_i' J_i'} | \left[\frac{1}{\omega - L_1 - n_2 \langle m(\omega) \rangle_B} \right] | \mathrm{v_f J_f v_i J_i} \rangle\rangle \quad [3.26]$$

where $[\mathrm{J}_\alpha] = 2\mathrm{J}_\alpha + 1$.

The matrix elements of the operator $\frac{1}{\omega - L_1 - n_2 \langle m(\omega) \rangle_B}$ can be extracted from the inversión of the matrix constructed on the operator's elements $\omega - L_1 - n_2 \langle m(\omega) \rangle_B$, which is a scalar operator. In this form, it is easy to see that the Liouville operator L_1 connected to the active molecule's Hamiltonian operator H_1 leads to the frequencies $\omega - \omega_{\mathrm{v_f J_f v_i J_i}} \delta_{\mathrm{v_f v_f'}} \delta_{\mathrm{v_i v_i'}} \delta_{\mathrm{J_f J_f'}} \delta_{\mathrm{J_i J_i'}}$. The calculation of the operator's elements $\omega - L_1 - n_2 \langle m(\omega) \rangle_B$ finally involves essentially that of the relaxation matrix $n_2 \langle m(\omega) \rangle_B$, which contains the dynamics of the collisions process.

3.2.4. *Semiclassical context for calculating the relaxation matrix*

In a semiclassical approach, only the rotation and vibration degrees of freedom are quantified. This approximation is valid if we can define a trajectory for the interacting molecules, which means that the dimension of the waves packet associated with the movement of each molecule is negligible in comparison to the range of molecular interaction forces. Within the framework of this approximation, the average $\langle m(\omega) \rangle_B$ of the operator $m(\omega)$, on the states of the bath, is expressed according to the trace on the perturbing molecule's vibration–rotation states and of the classical average of the set of the different trajectories.

In a coupling diagram like that of the active molecule, it is written that $\mathrm{J_2 - J_2' = K_2}$, in which $\mathrm{J_2}$ and $\mathrm{J_2'}$ designate the perturbing molecule's rotational angular momenta in the initial state and in the final state, respectively. The density operator's symmetry property (with spherical symmetry) impose $K_2 = 0$ so that in the expression of $n_2 \langle m(\omega) \rangle_B$ in equation [3.21], one has $\langle\langle 2'2' | = \langle\langle v'_2 J'_2 v'_2 J'_2 0 |$, $| 22 \rangle\rangle = | v_2 J_2 v_2 J_2 0 \rangle\rangle$ and $| 2 \rangle = | v_2 J_2 \rangle$. Finally, after simplifying by applying the Wigner–Eckart theorem and the properties of the Clebsch–Gordan coefficients, the matrix elements of the operator $n_2 \langle m(\omega) \rangle_B$ are written as:

$$\left\langle v'_f J'_f v'_i J'_i \left\| n_2 \langle m(\omega) \rangle_B \right\| v_f J_f v_i J_i \right\rangle = n_2 \sum_{v_2 J_2 m_2} \langle v_2 J_2 m_2 | \rho_2(0) | v_2 J_2 m_2 \rangle \times$$
$$\left\langle \frac{-(1)^{J_f + J'_f}}{\sqrt{[J_i][J'_i]}} \sum_{\substack{m_i m_f \\ m'_i m'_f}} \sum_{\substack{v'_2 J'_2 \\ m'_2}} C(J'_f J J'_i; m'_f M m'_i) C(J_f J J_i; m_f M m_i) \langle\langle f'2'i'2' | m(\omega) | f2i2 \rangle\rangle \right\rangle_{av} \quad [3.27]$$

with: $[J_\alpha] = 2J_\alpha + 1$;

$$\langle\langle f'2'i'2' | m(\omega) | f2i2 \rangle\rangle =$$
$$\langle\langle v'_f J'_f m'_f v'_2 J'_2 m'_2 v'_i J'_i m'_i v'_2 J'_2 m'_2 | m(\omega) | v_f J_f m_f v_2 J_2 m_2 v_i J_i m_i v_2 J_2 m_2 \rangle\rangle$$

and where $\langle \Theta_{ij} \rangle_{av}$ designates the classical average of the matrix element indexed by ij of the operator Θ. This average is evaluated using the characteristic parameters of a collision in space between two molecules, i.e. the relative velocity v of the two molecules before they collide, the impact parameter b, the instant of closest approach t_0 and Euler's angles (α_i, β_i, γ_i) defining the molecule's orientation i (active molecule or perturbing molecule).

The average on Euler's angles is given by:

$$\langle \Theta_{ij} \rangle_{\alpha\beta\gamma} = \frac{1}{8\pi^2} \int_0^{2\pi} d\alpha \int_0^{\pi} \sin\beta \, d\beta \int_0^{2\pi} \Theta_{ij}(v, b, t_0, \alpha, \beta, \gamma) d\gamma \quad [3.28]$$

and represents an average on all possible orientations of the trajectories. This average does not interfere given that the operator $m(\omega)$ is invariant by rotation. For the classical average, it is written as:

$$\left\langle \Theta_{ij} \right\rangle_{av} = \int_0^{\infty} v\mathrm{f}(v)\,dv \int_0^{\infty} 2\pi \mathrm{b}\, d\mathrm{b} \int_{-\infty}^{+\infty} \Theta_{ij}(v,\mathrm{b},t_0)\,dt_0 \qquad [3.29]$$

where f(v) is the Maxwell three-dimensional velocity distribution, in which the expression is $\mathrm{f}(v) = 4\pi\left(\frac{\mu}{2\pi k_B T}\right)^{3/2} v^2 \exp\text{-}\left(\frac{\mu v^2}{2k_B T}\right)$ and in which μ is the reduced mass of the collision partners.

Since the operator $m(\omega)$ is the solution to the Lippman–Schwinger [LIP 50] equation, which can be expressed as follows:

$$m(\omega) = (\omega - L_0)\left(\frac{1}{\omega - L_0 - L_{12}} - \frac{1}{\omega - L_0}\right)(\omega - L_0) \qquad [3.30]$$

It is then possible according to the following considerations:

– its expansion in the space of time;

– the impact approximation (instantaneous binary collisions);

– the hypothesis that the gap between the two neighboring lines is greater than the inverse of the average duration t_c during which the two partner molecules of a collision are subjected to the molecular interaction potential, meaning that: $2\pi\Delta\omega = 2\pi(\omega_{i'f'} - \omega_{if}) >> \frac{1}{t_c}$, to expand the reduced matrix element of the operator $\langle m(\omega)\rangle_B$ of the Liouville space so as to obtain its expression in the Hilbert space as a function of the time evolution operator.

If the duration of the collision tc is negligible compared to the time separating two successive collisions (instantaneous binary collision: impact approximation) and if at the moment of a collision, the molecule no longer preserves any memory of the previous collision (Markov process), it is shown that m(ω) does not vary significantly near the center of an absorption or emission line so that in the frequency domain corresponding to

$\Delta\omega = \omega - \omega_{if} << \frac{1}{t_c}$, m(ω) can be considered as constant and independent of the frequency.

The operator is then expressed according to the diffusion operator *S*, which describes the evolution of the system's states under the influence of the perturbation $V_{12}(t)$. This operator is symbolically written as follows:

$$S = \text{T}\exp\left(-\frac{i}{\hbar}\int_{-\infty}^{+\infty}\tilde{V}_{12}(t)\,dt\right) \qquad [3.31]$$

where T designates the Dyson chronological order operator [DYS 49] and $\tilde{V}_{12}(t) = \exp\left(\frac{i}{\hbar}(H_1 + H_2)t\right)V_{12}\exp\left(-\frac{i}{\hbar}(H_1 + H_2)t\right)$ represents the interaction potential in the interaction representation.

The expansion of the operator $\langle m(\omega)\rangle_B$, initially expressed in Liouville space, is then written in Hilbert space as:

$$\begin{aligned}&\left\langle \mathrm{v}'_f\mathrm{J}'_f\mathrm{v}'_i\mathrm{J}'_i\right\| n_2\langle m(\omega)\rangle_B \left\| \mathrm{v}_f\mathrm{J}_f\mathrm{v}_i\mathrm{J}_i\right\rangle = -in_2\sum_{\mathrm{v}_2\mathrm{J}_2\mathrm{m}_2}\langle \mathrm{v}_2\mathrm{J}_2\mathrm{m}_2|\rho_2(0)|\mathrm{v}_2\mathrm{J}_2\mathrm{m}_2\rangle\times\\ &\int_0^\infty 2\pi\mathrm{b}\,d\mathrm{b}\int_0^\infty v\mathrm{f}(v)dv\frac{-(1)^{\mathrm{J}_f+\mathrm{J}'_f}}{\sqrt{[\mathrm{J}_i][\mathrm{J}'_i]}}\sum_{\substack{\mathrm{m}_i\mathrm{m}_f\mathrm{M}\\ \mathrm{m}'_i\mathrm{m}'_f}}\sum_{\substack{\mathrm{v}'_2\mathrm{J}'_2\\ \mathrm{m}'_2\mathrm{m}_2}}\mathrm{C}(\mathrm{J}'_f\mathrm{J}\mathrm{J}'_i;\mathrm{m}'_f\mathrm{M}\mathrm{m}'_i)\mathrm{C}(\mathrm{J}_f\mathrm{J}\mathrm{J}_i;\mathrm{m}_f\mathrm{M}\mathrm{m}_i)\\ &\begin{pmatrix}\delta_{\mathrm{ff}'}\delta_{\mathrm{ii}'}\delta_{22'} - \langle \mathrm{v}'_f\mathrm{J}'_f\mathrm{m}'_f\mathrm{v}'_2\mathrm{J}'_2\mathrm{m}'_2|S|\mathrm{v}_f\mathrm{J}_f\mathrm{m}_f\mathrm{v}_2\mathrm{J}_2\mathrm{m}_2\rangle\times\\ \langle \mathrm{v}_i\mathrm{J}_i\mathrm{m}_i\mathrm{v}_2\mathrm{J}_2\mathrm{m}_2|S^+|\mathrm{v}'_i\mathrm{J}'_i\mathrm{m}'_i\mathrm{v}'_2\mathrm{J}'_2\mathrm{m}'_2\rangle\end{pmatrix}\end{aligned} \qquad [3.32]$$

The isolated lines or for which the interferences are negligible, only involve the diagonal elements of the relaxation matrix *m*(ω), which is constant and independent of the frequency being written in the form $n_2\langle m(\omega)\rangle_B = n_2\langle m(0)\rangle_B = \Delta - i\Gamma$, in which Δ and Γ are two hermitic operators in Liouville space. From equation [3.27], one extracts:

$$\begin{aligned}&\left\langle \mathrm{v}_f\mathrm{J}_f\mathrm{v}_i\mathrm{J}_i\right\| n_2\langle m(0)\rangle_B\left\| \mathrm{v}_f\mathrm{J}_f\mathrm{v}_i\mathrm{J}_i\right\rangle\\ &= \left\langle \mathrm{v}_f\mathrm{J}_f\mathrm{v}_i\mathrm{J}_i\right\|\Delta\left\| \mathrm{v}_f\mathrm{J}_f\mathrm{v}_i\mathrm{J}_i\right\rangle - i\left\langle \mathrm{v}_f\mathrm{J}_f\mathrm{v}_i\mathrm{J}_i\right\|\Gamma\left\| \mathrm{v}_f\mathrm{J}_f\mathrm{v}_i\mathrm{J}_i\right\rangle = \delta_{\mathrm{fi}} - i\gamma_{\mathrm{fi}}\end{aligned} \qquad [3.33]$$

where δ_{fi} and γ_{fi} are the eigenvalues of the operators Δ and Γ.

As the matrix of the operator $\Theta^{-1} = \omega - L_1 - n_2\langle m(\omega)\rangle_B$ is diagonal, it can be directly inverted to obtain the matrix elements of $\Theta = \left(\omega - L_1 - n_2\langle m(\omega)\rangle_B\right)^{-1}$ thus allowing the calculation of the spectral shape based on equation [3.26].

3.2.5. *Broadening parameter according to the diffusion operator*

In the case of the broadening of isolated lines, the non-diagonal matrix elements of the relaxation operator are negligible. The spectral line shape calculated from the matrix elements of $\Theta = \left(\omega - L_1 - n_2\langle m(\omega)\rangle_B\right)^{-1}$ finally leads to the following expression:

$$
\begin{aligned}
I(\omega) = \frac{1}{\pi[\mathrm{J}]} \sum_{\substack{\mathrm{v_iJ_im_i\,v_fJ_fm_f} \\ \mathrm{v_i'J_i'v_f'J_f'}}} & [\mathrm{J_f}] \left|\langle \mathrm{v_fJ_f} \| \mathrm{X^J} \| \mathrm{v_iJ_i}\rangle\right|^2 \\
\times & \left[\langle \mathrm{v_iJ_im_i} | \rho_1(0) | \mathrm{v_iJ_im_i}\rangle \frac{\gamma_{\mathrm{fi}}}{(\omega - \omega_{\mathrm{fi}} - \delta_{\mathrm{fi}})^2 + \gamma_{\mathrm{fi}}^2} + \right. \\
& \left. \langle \mathrm{v_fJ_fm_f} | \rho_1(0) | \mathrm{v_fJ_fm_f}\rangle \frac{\gamma_{\mathrm{fi}}}{(\omega - \omega_{\mathrm{fi}} + \delta_{\mathrm{fi}})^2 + \gamma_{\mathrm{fi}}^2} \right]
\end{aligned}
\qquad [3.34]
$$

which corresponds to a Van Vleck–Weisskopf line shape [VAN 45]. The two Lorentzian lines that appear in the sum between the square brackets correspond to the positive and negative frequencies, respectively. This is equivalent to considering an absorption line as being made up of a line peaked at an angular frequency ω_{fi} and of the distant wing of a second line with an angular frequency peaked at $-\omega_{\mathrm{fi}}$. In the infrared, the contribution of the latter is negligible compared to that of the first and only the Lorentz spectral line shape of the first term is kept and in which δ_{fi} and γ_{fi}, respectively, represent the shift and the half-width at half-maximum of the spectral line of angular frequency ω_{fi}.

The half-width at half-maximum γ_{fi} is calculated from:

$$\gamma_{\text{fi}} = n_2 \operatorname{Re} \sum_{\text{v}_2 \text{J}_2} \rho_{2\text{v}_2 \text{J}_2} \int_0^{\infty} 2\pi \text{b}\, d\text{b} \int_0^{\infty} v\text{f}(v) S(\text{b},v) dv \qquad [3.35]$$

in which S(b,v) is the differential cross-section whose expression is given by:

$$\begin{aligned} S(\text{b},v) = & \frac{1}{[\text{J}_2][\text{J}_\text{i}]} \sum_{\substack{\text{m}_\text{i}\text{m}_\text{f} \\ \text{m}'_\text{i}\text{m}'_\text{f}}} \sum_{\substack{\text{v}'_2\text{J}'_2 \\ \text{m}_2\text{m}'_2}} \text{C}(\text{J}_\text{f}\text{JJ}_\text{i};\text{m}'_\text{f}\text{Mm}'_\text{i})\text{C}(\text{J}_\text{f}\text{JJ}_\text{i};\text{m}_\text{f}\text{Mm}_\text{i}) \\ & \times \Big[\delta_{\text{m}_\text{f}\text{m}'_\text{f}} \delta_{\text{m}_\text{i}\text{m}'_\text{i}} \delta_{\text{m}_2\text{m}'_2} \delta_{\text{J}_2\text{J}'_2} \delta_{\text{v}_2\text{v}'_2} - \langle \text{v}_\text{f}\text{J}_\text{f}\text{m}_\text{f}\text{v}_2\text{J}_2\text{m}_2 | S | \text{v}_\text{f}\text{J}_\text{f}\text{m}_\text{f}\text{v}'_2\text{J}'_2\text{m}'_2 \rangle \\ & \qquad \times \langle \text{v}_\text{i}\text{J}_\text{i}\text{m}_\text{i}\text{v}_2\text{J}_2\text{m}_2 | S^+ | \text{v}_\text{i}\text{J}_\text{i}\text{m}_\text{i}\text{v}'_2\text{J}'_2\text{m}'_2 \rangle \Big] \end{aligned} \qquad [3.36]$$

and in which:

$$\rho_{2\text{v}_2\text{J}_2} = [\text{J}_2] \sum_{\text{v}_2\text{J}_2} \langle \text{v}_2\text{J}_2\text{m}_2 | \rho_2(0) | \text{v}_2\text{J}_2\text{m}_2 \rangle \qquad [3.37]$$

3.2.6. *Calculation of the differential cross-section S(b, v)*

In the theoretical model proposed by Anderson [AND 49] and generalized by Tsao and Curnutte [TSA 62], the non-commutation effects of interaction potential are assumed to be negligible at all orders of approximation. The perturbation theory is applied up to the second order of the expansion of the diffusion operator S:

$$S = 1 - \frac{i}{\hbar} \int_{-\infty}^{+\infty} \tilde{V}_{12}(t)\, dt + \frac{1}{2} \left(\frac{-i}{\hbar} \right)^2 \left| \int_{-\infty}^{+\infty} \tilde{V}_{12}(t)\, dt \right|^2 \qquad [3.38]$$

The differential cross-section S(b, v) is then expressed as a sum of three terms: $S(\text{b},v) = S_0(\text{b},v) + S_1(\text{b},v) + S_2(\text{b},v)$.

The zeroth-order term gives a null contribution; the first-order term is purely imaginary and the second-order term is real (when the non-commutation effects of the interaction potential are ignored). The broadening parameter γ_{fi} is then expressed in terms of $S_2(\text{b},v)$ that, following the notation introduced by Anderson [AND 49], can be put in the following form:

$$S_2(\mathrm{b},v) = S_2(\mathrm{b},v)_{\mathrm{outer,i}} + S_2(\mathrm{b},v)_{\mathrm{middle}} + S_2(\mathrm{b},v)_{\mathrm{outer,f}} \qquad [3.39]$$

Each of the terms are expressed according to the matrix elements of the collision operator P defined by:

$$\mathrm{P} = -\frac{i}{\hbar}\int_{-\infty}^{+\infty} \tilde{V}_{12}(t)\,dt \qquad [3.40]$$

The expressions of these different terms are given in the Appendix (section 3.6).

Robert and Bonamy [ROB 79] apply the linked cluster theorem and use the cumulant expansion [BLO 58, BRU 55, GOL 57, KUB 62] to calculate the differential cross-section. This theorem can easily be applied by expressing the relaxation matrix in the form $n_2\langle m(\omega)\rangle_B = n_2\langle m(0)\rangle_B = -in_2\langle 1-\hat{S}\rangle_B$ [FAN 63] in which $\hat{S}$ is the diffusion operator in the Liouville space and defined by $\hat{S} = (S\otimes \mathrm{I}^+)(\mathrm{I}\otimes S^+) = S\otimes S^+ = SS^+$. The matrix element of the operator can then be written in the following form:

$$\begin{aligned}&\langle\langle \mathrm{v}_f'\mathrm{J}_f'\mathrm{v}_i'\mathrm{J}_i'\mathrm{J}|n_2\langle m(0)\rangle_B|\mathrm{v}_f\mathrm{J}_f\mathrm{v}_i\mathrm{J}_i\mathrm{J}\rangle\rangle\\ &\quad= -in_2\left\langle \delta_{\mathrm{v}_f'\mathrm{v}_f}\delta_{\mathrm{J}_f'\mathrm{J}_f}\delta_{\mathrm{v}_i'\mathrm{v}_i}\delta_{\mathrm{J}_i'\mathrm{J}_i} - \langle\langle \mathrm{v}_f'\mathrm{J}_f'\mathrm{v}_i'\mathrm{J}_i'\mathrm{J}|\hat{S}|\mathrm{v}_f\mathrm{J}_f\mathrm{v}_i\mathrm{J}_i\mathrm{J}\rangle\rangle\right\rangle_{\mathrm{av}}\end{aligned} \qquad [3.41]$$

The average $\langle\ldots\rangle_{\mathrm{av}} = \langle\ldots\rangle_{\mathrm{bv2}}$ designates the average on the impact parameter b, the relative velocity v and the quantum states of the perturbator 2 such as:

$$\langle\hat{S}\rangle_2 = \sum_{\substack{\mathrm{v}_2\mathrm{J}_2\\ \mathrm{v}_2'\mathrm{J}_2'}} \langle\langle \mathrm{v}_2'\mathrm{J}_2'\mathrm{v}_2'\mathrm{J}_2'\mathrm{J}|\hat{S}|\mathrm{v}_2\mathrm{J}_2\mathrm{v}_2\mathrm{J}_2\mathrm{J}\rangle\rangle\langle \mathrm{v}_2\mathrm{J}_2|\rho_2(0)|\mathrm{v}_2\mathrm{J}_2\rangle \qquad [3.42]$$

After separating the averages on the impact parameter b and the relative velocity v, the differential cross-section is calculated from the following matrix element:

$$\langle\langle \mathrm{v}_f\mathrm{J}_f\mathrm{v}_2'\mathrm{J}_2'\mathrm{v}_i\mathrm{J}_i\mathrm{v}_2'\mathrm{J}_2'\mathrm{J}|\hat{S}|\mathrm{v}_f\mathrm{J}_f\mathrm{v}_2\mathrm{J}_2\mathrm{v}_i\mathrm{J}_i\mathrm{v}_2\mathrm{J}_2\mathrm{J}\rangle\rangle = \langle\langle \mathrm{f2'i2'}|\hat{S}|\mathrm{f2i2}\rangle\rangle \qquad [3.43]$$

with

$$\left|v_f J_f v_2 J_2 v_i J_i v_2 J_2 J\right\rangle\rangle = \sum_{m_i m_f m_2} (-1)^{m_i + J_2} C(J_f J_i J; m_i m_f m) \frac{1}{[J_2]} \left|v_f J_f m_f v_2 J_2 m_2\right\rangle \left|v_i J_i m_i v_2 J_2 m_2\right\rangle^{+} \quad [3.44]$$

which is a vector of the standard basis constructed from the composition of angular momenta K_1 and K_2 [COH 73, MES 64]. These angular momenta are associated with the partner molecules of a collision such as $K = K_1 + K_2$ with $K_2 = 0$, which finally leads to $K = J$.

Under the assumption that the cumulant expansion is valid in the Liouville space to evaluate the statistical and quantum averages, the matrix elements $\langle\langle f2'i2'|\hat{S}|f2i2\rangle\rangle$ are developed in the form of a product of two terms [ROB 79] $\langle\langle f2'i2'|\hat{S}|f2i2\rangle\rangle = \langle\langle f2'i2'|\hat{S}^{(L)}|f2i2\rangle\rangle$ $\exp\left(\langle\langle f2i2|\hat{S}^{(C)}|f2i2\rangle\rangle\right)$, in which the exponents L and C are, respectively, related to the contributions of the linked diagrams and the isolated connected diagrams.

By limiting this expansion to the second-order diagrams, it can be written that:

$$\begin{aligned}\sum_{2'}\langle\langle f2'i2'|\hat{S}|f2i2\rangle\rangle = &\left[\sum_{2'}\delta_{22'} - \frac{i}{\hbar}\int_{-\infty}^{+\infty} dt \langle\langle f2'i2'|\tilde{V}_{12}(t)|f2i2\rangle\rangle_L\right. \\ &\left.-\frac{1}{\hbar^2}\int_{-\infty}^{+\infty} dt \int_{-\infty}^{t} dt' \langle\langle f2'i2'|\tilde{V}_{12}(t)\tilde{V}_{12}(t')|f2i2\rangle\rangle_L\right] \exp\left(\frac{-i}{\hbar}\int_{-\infty}^{+\infty} dt \langle\langle f2i2|\tilde{V}_{12}(t)|f2i2\rangle\rangle_C\right. \\ &\left.-\frac{1}{\hbar^2}\int_{-\infty}^{+\infty} dt \int_{-\infty}^{t} dt' \langle\langle f2i2|\tilde{V}_{12}(t)\tilde{V}_{12}(t')|f2i2\rangle\rangle_C\right)\end{aligned} \quad [3.45]$$

Given the zero contribution, in the first order, of the bound states $\left(S_1^{(L)}\right) = 0$, the differential cross-section (equation [3.34]) calculated from this expansion is expressed as:

$$S(b,v) = 1 - \left[1 - S_2^{(L)}(b,v)\right]\exp\left(-S_2^{(C)}(b,v)\right)\exp\left(-iS_1^{(C)}(b,v)\right) \quad [3.46]$$

with

$$S_1^{(C)}(b,v) = S_{1,f}(b,v) - S_{1,i}(b,v) \quad [3.47a]$$

$$S_2^{(L)}(b,v) = S_2^{ND}(b,v)_{middle} \quad [3.47b]$$

$$S_2^{(C)}(b,v) = S_2(b,v)_{outer,i} + S_2(b,v)_{outer,f} + S_2^{D}(b,v)_{middle} + iS_2^{(C')}(b,v) \quad [3.47c]$$

The exponents D and ND, respectively, designate the diagonal and non-diagonal contributions on the states of the perturbator and $S_2^{(C')}(b,v)$ is the imaginary part of $S_2^{(C)}(b,v)$. The expressions of $S_2(b,v)_{outer,i}$, $S_2(b,v)_{middle}$, $S_2(b,v)_{outer,f}$ and $S_2^{(C)}(b,v)$ are given in the Appendix (section 3.6).

The broadening parameter γ_{fi}^0 can also be calculated from the following expression:

$$\gamma_{fi}^0(RB) = \frac{n_2}{2\pi c}\sum_{v_2J_2}\rho_{2v_2J_2}\int_0^\infty 2\pi b\, db\int_0^\infty dv\, vf(v)\Big[1-\big(1-S_2^{ND}(b,v)_{middle}\big)\times \Big(\exp-\big(S_2(b,v)_{outer,i}+S_2(b,v)_{outer,f}+S_2^{D}(b,v)_{middle}\big)\Big)\cos\Big(S_1^{(C)}(b,v)+S_2^{(C')}(b,v)\Big)\Big] \quad [3.48]$$

One can also replace the term $1\text{-}S_2^{(L)}(b,v)$ by $\exp\left(\text{-}S_2^{(L)}(b,v)\right)$, which is valid in the case of Q branch lines in isotropic Raman diffusion.

By considering that the cumulant expansion is applied to the vectors of the Hilbert space and not directly to the vectors of Liouville space, which are Hilbert space operators, Ma *et al.* [MA 07] express equation [3.48] differently, by calculating the average on the states of the bath before applying the cumulant expansion. The average on the states of the bath appears inside the expansion and we have:

$$\gamma_{fi}^0(MTB) = \frac{n_2}{2\pi c}\int_0^\infty 2\pi b\, db\int_0^\infty dv\, vf(v)\Big[1\text{-}\Big(1-\big\langle S_2^{ND}(b,v)_{middle}\big\rangle_{v_2J_2}\Big)\times \Big(\exp-\Big(\big\langle S_2(b,v)_{outer,i}\big\rangle_{v_2J_2}+\big\langle S_2(b,v)_{outer,f}\big\rangle_{v_2J_2}+\big\langle S_2^{D}(b,v)_{middle}\big\rangle_{v_2J_2}\Big)\Big)\times \cos\Big(\big\langle S_1^{(C)}(b,v)\big\rangle_{v_2J_2}+\big\langle S_2^{(C')}(b,v)\big\rangle_{v_2J_2}\Big)\Big] \quad [3.49]$$

Similarly, the expressions of the shifts are different, given the different application of the cumulant expansion:

$$\delta^0_{\text{if}}(\text{RB}) = \frac{n_2}{2\pi c}\sum_{\text{v}_2\text{J}_2}\rho_{2\text{v}_2\text{J}_2}\int_0^\infty 2\pi\text{b}\, d\text{b}\int_0^\infty dv\, v\text{f}(v)\Big[\big(1 - S_2^{\text{ND}}(\text{b},v)_{\text{middle}}\big)\times \\ \big(\exp - \big(S_2(\text{b},v)_{\text{outer,i}} + S_2(\text{b},v)_{\text{outer,f}} + S_2^{\text{D}}(\text{b},v)_{\text{middle}}\big)\big)\sin\big(S_1^{(\text{C})}(\text{b},v) + S_2^{(\text{C}')}(\text{b},v)\big)\Big] \quad [3.50]$$

$$\delta^0_{\text{if}}(MTB) = \frac{n_2}{2\pi c}\int_0^\infty 2\pi\text{b}\, d\text{b}\int_0^\infty dv\, v\text{f}(v)\Big[\Big(1 - \big\langle S_2^{\text{ND}}(\text{b},v)_{\text{middle}}\big\rangle_{\text{v}_2\text{J}_2}\Big)\times \\ \Big(\exp - \Big(\big\langle S_2(\text{b},v)_{\text{outer,i}}\big\rangle_{\text{v}_2\text{J}_2} + \big\langle S_2(\text{b},v)_{\text{outer,f}}\big\rangle_{\text{v}_2\text{J}_2} + \big\langle S_2^{\text{D}}(\text{b},v)_{\text{middle}}\big\rangle_{\text{v}_2\text{J}_2}\Big)\Big)\times \\ \sin\Big(\big\langle S_1^{(\text{C})}(\text{b},v)\big\rangle_{\text{v}_2\text{J}_2} + \big\langle S_2^{(\text{C}')}(\text{b},v)\big\rangle_{\text{v}_2\text{J}_2}\Big)\Big] \quad [3.51]$$

in which $\langle\Theta\rangle_{\text{v}_2\text{J}_2}$ is given by:

$$\langle\Theta\rangle_{\text{v}_2\text{J}_2} = \sum_{\text{v}_2\text{J}_2}\Theta\rho_{2\text{v}_2\text{J}_2} \quad [3.52]$$

When the interactions between the active molecule and the perturbing molecule are weak, both expressions lead to values that are barely different. When the interactions are stronger, the formula given by Robert and Bonamy must be modified to the benefit of that of Ma *et al.* Gamache *et al.* [ANT 06, GAM 97, GAM 98, LYN 96, NES 94] pushed the expansion of the interactions to higher orders when applying the Robert and Bonamy model. In particular, they show that in the CRB model ("Complex Robert Bonamy Formalism") the effects of the imaginary terms must be considered in the differential cross-section for calculating the widths and shifts of the lines. By comparing results calculated with the CRB and MTB [ANT 06] methods, they show that there is better correspondence between the simulation results and the experimentation with CRB than with MTB.

3.2.7. *Interaction potential energy*

To account for all the short- and mid-range repulsive interactions and dispersion forces, a 12-6 Lennard–Jones type intermolecular interaction [HIR 84] is very often used (Figure 3.1) such as:

$$V_a(\mathrm{R}) = 4\varepsilon\left(\left(\frac{\sigma}{\mathrm{R}}\right)^{12} - \left(\frac{\sigma}{\mathrm{R}}\right)^{6}\right) \qquad [3.53a]$$

where ε is the depth of the potential wells and σ is the value of R for which the potential is cancelled out ($V_a(\sigma)=0$) and $\mathrm{R} = \frac{\sigma}{2^{1/6}}$ is the abscissa of the minimum of the potential. In the case where the repulsive interaction is particularly strong, the Buckingham type potential can be used:

$$V_a(\mathrm{R}) = \mathrm{B}e^{-b\mathrm{R}} - \frac{\mathrm{A}}{\mathrm{R}^6} \qquad [3.53b]$$

where b, A and B are parameters independent of R.

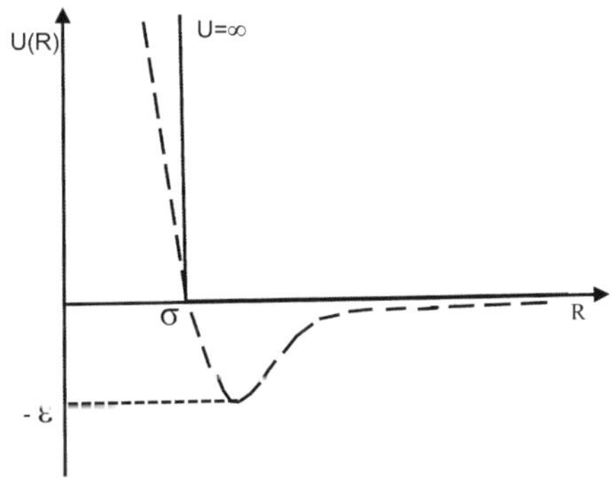

Figure 3.1. *Isotropic part of the potential: - - - - - Lennard-Jones Potential; ____ hard sphere potential; R: intermolecular distance; ε: depth of the potential wells; σ: value of R such as the potential is cancelled out ($V_a(\sigma) = 0$)*

These forms of potential, independent of the relative orientation of the interacting molecules can be used to determine the relative trajectory of the partner molecules of a collision in a semiclassical model. Different models have been proposed to represent the angular dependence of the short- to mid-range intermolecular interaction. One of the most used is the atom–atom potential, which is expressed as a superposition of the Lennard–Jones type (12-6) interactions between a pair of atoms, each of which is located on one of the partner molecules of a collision. The angular dependence according to the relative orientation of the molecules is expressed as the products of spherical harmonics.

In Robert and Bonamy's model, the interaction potential comprises a term made up of a superposition of a short-range atom–atom interaction of 12-6 Lennard–Jones type $V_a\left(\vec{R}(t)\right)$ between the partner molecules of a collision and a long-range electro-static interaction term $V_E\left(\vec{R}(t)\right)$.

$$V_a\left(\vec{R}(t)\right)=\sum_{i,j}\left(\frac{d_{ij}}{r_{1i,2j}^{12}}-\frac{e_{ij}}{r_{1i,2j}^{6}}\right) \qquad [3.54a]$$

where i and j designate the atoms of molecules 1 and 2, respectively, $r_{1i,2j}$ is the relative distance between the interacting atoms, and d_{ij} and e_{ij} are the energy parameters of the pair (i, j).

There are other forms for expressing the atom–atom potential. Mac Rury *et al.* [MAC 76] express this potential in the following form:

$$V_a\left(\vec{R}(t)\right)=\sum_{i,j}\varepsilon\left(\left(\frac{\sigma}{r_{1i,2j}}\right)^{12}-\left(\frac{\sigma}{r_{1i,2j}}\right)^{6}\right) \qquad [3.54b]$$

This form was used by Murthy *et al.* [MUR 80] to study dinitrogen N_2 with 4ε for the energy term instead of ε. This results in the correspondence: $d_{ij}=\varepsilon\sigma^{12}$ and $e_{ij}=\varepsilon\sigma^{6}$.

The atom–atom potential is most often used in the form proposed by Oobatake and Ooi [OOB 72] as a two-site model.

$$V_a\left(\vec{R}(t)\right)=\sum_{i,j}\varepsilon_{ij}\left(\left(\frac{\sigma_{ij}}{r_{i,j}}\right)^{12}-2\left(\frac{\sigma_{ij}}{r_{i,j}}\right)^{6}\right) \qquad [3.54c]$$

This model was also applied by Murthy *et al.* [MUR 81] to process the three-interaction sites case. In this case, one has the correspondence: $d_{ij}=\varepsilon\sigma^{12}$ and $e_{ij}=\varepsilon\sigma^{6}$.

It is generally necessary to apply the combination rules to determine the interaction parameters between two different atoms from the interaction parameters between two atoms of the same kind. The Berthelot and Lorentz rules lead to: $\sigma_{ij} = \left(\sigma_{ii} + \sigma_{jj}\right)/2$ and $\varepsilon_{ij} = \sqrt{\varepsilon_{ii}\varepsilon_{jj}}$. The rules proposed by Oobatake and Ooi lead to the expressions: $\sigma_{ij} = \left(\sigma_{ii} + \sigma_{jj}\right)/2$, $\mathrm{e}_{ij} = \sqrt{\mathrm{e}_{ii}\mathrm{e}_{jj}}$ and $\mathrm{d}_{ij} = \mathrm{e}_{ij}\,\sigma_{ij}^{6}/2$.

When V_{a} is expanded as a function of the intermolecular distance $\mathrm{R}(t)$, the intramolecular distances r_{1i} and r_{2j}, and the spherical harmonic products associated with each molecule in the mobile frame (whose axis z is directed along the intermolecular axis), the following expression is obtained:

$$V_{\mathrm{a}}\left(\vec{\mathrm{R}}(t)\right) = \sum_{i,j}\sum_{\mathrm{q}=0}^{\mathrm{n}}\sum_{\substack{l_1=0\\ l_2=0}}^{\mathrm{n}}\sum_{m_1,m_2=-\inf(l_1,l_2)}^{m_1,m_2=+\inf(l_1,l_2)}\left(\frac{{}^{\mathrm{q}}_{ij}\mathrm{D}_{l_1l_2}^{m_1m_2}\,\mathrm{d}_{ij}}{\mathrm{R}^{12+\mathrm{q}}} - \frac{{}^{\mathrm{q}}_{ij}\mathrm{E}_{l_1l_2}^{m_1m_2}\,\mathrm{e}_{ij}}{\mathrm{R}^{6+\mathrm{q}}}\right)Y_{l_1}^{m_1}(\theta_1,\phi_1)\,Y_{l_2}^{m_2}(\theta_2,\phi_2) \qquad [3.54d]$$

The coefficients D and E are given in [BOU 82].

The electrostatic interaction potential between two non-overlapping charge distributions can be written in the following form:

$$V_{\mathrm{E}}\left(\vec{\mathrm{R}}(t)\right) = \sum_{l_1,l_2}\frac{\mathrm{K}(l_1,l_2,l)}{\mathrm{R}^{l+1}}\sum_{m_1m_2,m}\mathrm{C}(l_1,l_2,l;m_1m_2,m)\,\mathrm{Q}_{l_1}^{m_1}\mathrm{Q}_{l_2}^{m_2}\,Y_l^{m*}\left(\vec{\Omega}\right) \qquad [3.55]$$

with $\mathrm{K}(l_1,l_2,l) = \frac{(-1)^{l_2}\,4\pi}{(2l+1)}\sqrt{\frac{4\pi(2l+1)!}{(2l_1+1)!(2l_2+1)!}}$ and $l = l_1 + l_2$ (non-overlapping charges hypothesis).

In this expression, $\mathrm{C}(l_1,l_2,l;m_1m_2,m)$ designates the Clebsch–Gordan coefficients (different from zero if $m = m_1+m_2$; $\mathrm{Q}_{l_i}^{m_i}$ is the component m of the multipole of the order l of the distribution of charges bound to the molecule i, defined by [GRA 68]:

$$\mathrm{Q}_{l_i}^{m_i} = \sum_{\mathrm{n}_i} e_{\mathrm{n}_i}\mathrm{r}_{\mathrm{n}_i}^{l_i}\,Y_{l_i}^{m_i}\left(\vec{\omega}_{\mathrm{n}_i}\right) \qquad [3.56]$$

$Y_l^m(\vec{\Omega})$ and $Y_{l_i}^{m_i}(\vec{\omega}_{n_i})$ are spherical harmonics; $\vec{\Omega}$ and $\vec{\omega}_{n_i}$ are unitary vectors, respectively, defining the position of the charge n_i of the molecule i and the direction of the intermolecular axis with respect to a fixed frame (Figure 3.2); R is the intermolecular distance.

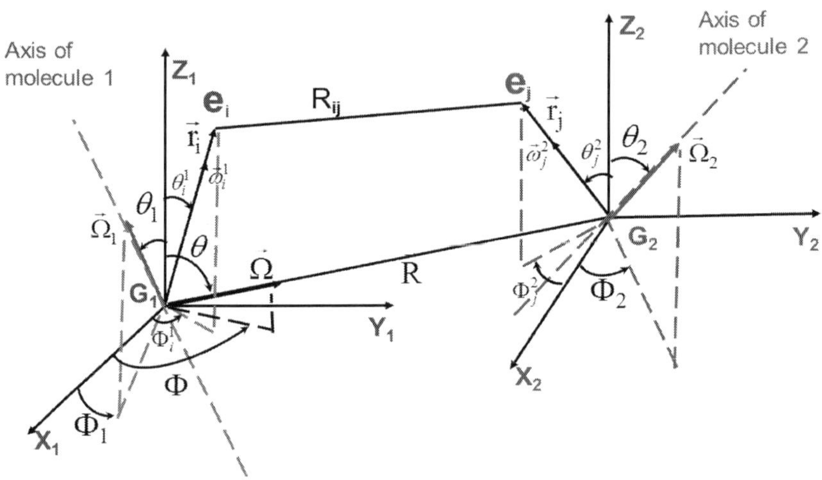

Figure 3.2. *The molecules 1 and 2 located in an axis system bound to the laboratory (fixed). For a color version of this figure, see www.iste.co.uk/dahoo/infrared.zip*

For a diatomic or linear molecule, the distribution of charges is of axial symmetry and one can write that: $Q_{l_i}^{m_i} = Q_{l_i} Y_{l_i}^{m_i}(\vec{\Omega}_i)$, with:

$$Q_{l_i} = \sum_{n_i} e_{n_i} r_{n_i}^{l_i} P_{l_i}(\cos\theta_{n_i}) \qquad [3.57]$$

where Q_{l_i} is the multipole of the order l of the molecule i; $P_{l_i}(\cos\theta_{n_i})$ is the Legendre polynomial of the order l, and $\vec{\Omega}_i$ is the unitary vector defining the orientation of the axis of symmetry of the molecule i.

If one chooses a reference system bound to the molecule, with the axis z along the intermolecular axis, the interaction potential is written in the form:

$$V_E\left(\vec{R}(t)\right)=\sum_{l_1,l_2}\frac{K(l_1,l_2,l)}{R^{l+1}}Q_{l_1}Q_{l_2}\sqrt{\frac{2l+1}{4\pi}}\sum_{m}C(l_1,l_2,l;m-m\ 0)Y_{l_1}^{m}\left(\vec{\Omega}_1\right)Y_{l_2}^{m}\left(\vec{\Omega}_2\right) \quad [3.58]$$

In the ATC model, the interaction potential is reduced to the electrostatic contribution only.

For the atom–atom potential, Robert and Bonamy use the form of equation [3.53a] to apply their theoretical model. Given all of the interaction forces, the intermolecular potential is expressed as a sum of different contributions such as: $V(t) = V_R(t) + V_E(t) + V_P(t)$ where the indices R, E and P relate to the repulsive, electrostatic and polarization interaction energies. This interaction potential is finally put in the following form [ROB 79]:

$$V\left(\vec{R}(t)\right)=4\pi\sum_{m=-\inf(l_1,l_2)}^{m=+\inf(l_1,l_2)}U_{l_1l_2m}\left(R(t)\right)Y_{l_1}^{m}\left(\vec{\Omega}_1\right)Y_{l_2}^{-m}\left(\vec{\Omega}_2\right) \quad [3.59]$$

From this expression, it results that, regardless of the interaction type considered, the matrix elements of the collision operator involve the expressions of the type: $\langle J_1M_1J_2M_2|Y_{l_1}^{m_1}(\vec{\Omega}_1)Y_{l_2}^{m_2}(\vec{\Omega}_2)|J_1M_1J_2M_2\rangle$. Given that $S_2(b,v)$ is a quadratic function of the interaction potential, it is necessary to account for the interferences between interactions of different kinds, when the same product of spherical harmonics appears in the expansion of the different contributions, i.e. electrostatic and atom–atom potential.

3.2.8. *Relative trajectory of the molecules*

In semiclassical formalism, the trajectory of the partner molecules of a collision is driven by the isotropic part U(R) of the interaction potential between the absorbing molecule and a molecule of the bath and which therefore does not depend on the relative orientation of the two molecules. To determine this trajectory, it is necessary to solve the equations of motion for both interacting molecules.

A two-body problem under mutual interaction can be reduced to a problem with a reduced mass μ moving in a central field U(R). The fictitious particle's movement occurs in a plane that is perpendicular to its orbital angular momentum $\vec{L}$, which is a constant of motion.

From the energy and angular momentum conservation rules, one finds that:

$$E = \frac{\mu \dot{R}^2}{2} + \frac{L^2}{2\mu R^2} + U(R) \qquad [3.60a]$$

with:

$$\frac{dR}{dt} = \dot{R} = \sqrt{\left[\frac{2}{\mu}\left(E\text{-}U(R)\right) - \frac{L^2}{\mu^2 R^2}\right]} \qquad [3.60b]$$

and

$$\frac{d\theta(t)}{dt} = \dot{\theta}(t) = \frac{L^2}{\mu R^2} \qquad [3.60c]$$

in which R(t) and θ(t) are the variables defining the trajectory in the collision's plane (Figure 3.3).

The equality $\dot{R}(t)=0$ defines a cusp of the trajectory, where the function R(t) at first increases and then decreases or vice versa. It defines the minimal distance of approach R_c. Given that the field is central, the two asymptotes on the trajectory cut the straight line passing through the point of closest distance R_c, under the same angle (Figure 3.3) $\chi = |\pi - 2\theta(0)|$ in which $\theta(0)$ is the angle that $\overrightarrow{OM}$ makes with the particle's incident direction.

When a collision is characterized by the impact parameter b, and by the velocity v of the particle at infinity, it can be written that $E = \frac{\mu v^2}{2}$ and that $L = \mu v b$, so that the differential equations [3.60b] and [3.60c] become:

$$\frac{dR}{dt} = v\sqrt{\left[1 - \frac{U(R)}{E} - \frac{b^2}{R^2}\right]} \qquad [3.61a]$$

and

$$R^2 \frac{d\theta(t)}{dt} = vb \qquad [3.61b]$$

From this, it results that:

$$\frac{R_c}{b} = \left[1 - \frac{U(R)}{E}\right]^{-1/2} \qquad [3.62]$$

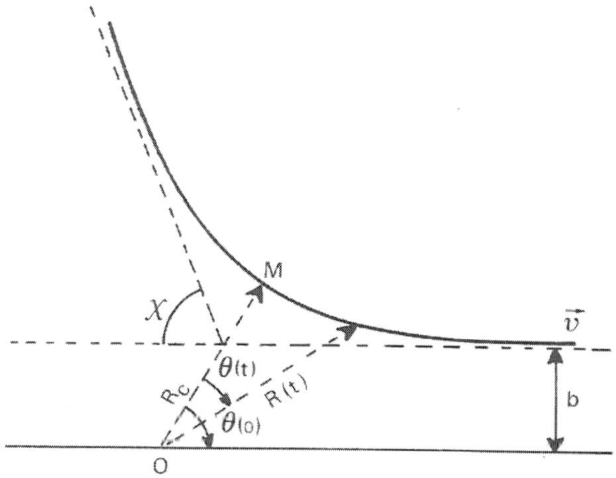

Figure 3.3. *Relative trajectory of the molecules; OM: distance of the minimum approach; v: velocity of the fictitious particle at infinity; X: deviation angle*

In the ATC theoretical model, the relative trajectory of the two colliding molecules is assumed to be a straight line and the velocity to be constant. In the framework of this approximation, this means that a hard sphere potential (Figure 3.2) is used for the isotropic part of the potential (U(R) << E) as long as the impact parameter is greater than the molecular diameter of the absorbing molecule.

In these conditions, one has: $R(0) = R_c = b$ and consequently:

$$R(t) = \sqrt{\left(b^2 + (vt)^2\right)} \qquad [3.63a]$$

and:

$$\theta(t) = \text{Arc}\sin\left(\frac{vt}{\mathrm{R}(t)}\right) = \text{Arc}\cos\left(\frac{\mathrm{b}}{\mathrm{R}(t)}\right) \qquad [3.63b]$$

The hypothesis of a rectilinear trajectory is valid when the intermolecular forces are negligible, meaning that the kinetic energy is preponderant on the potential energy. When the molecules draw near one another during a collision, the effect of the intermolecular forces modifies the initial rectilinear trajectory of the molecules and this modification is even more significant when the impact parameter is small.

From the characteristic parameters of a collision, i.e. the impact parameter b and the velocity v, it can be shown that as the interaction potential is expressed according to the higher negative powers of R(t) (equation [3.63a]), the contribution of the nonlinear part of the trajectory through R(t) is only significant around the closest distance of approach $\mathrm{R_c}$ and the relative velocity v_c at this point. Just as Tipping and Herman [TIP 70] have done, the impact parameter b and the velocity v in equation [3.63a] can be replaced by the closest distance of approach $\mathrm{R_c}$ and the relative velocity v_c at this point such as: $\mathrm{R}(t) = \sqrt{\left(\mathrm{R_c^2} + v_c^2 t^2\right)}$ for the distance between the molecules. Taking the conservation of the angular momentum into account, which implies that $v_c\mathrm{R_c} = v\mathrm{b}$, the expression of $\mathrm{R}(t)$ is finally given in the form:

$$\mathrm{R}(t) = \sqrt{\left(\mathrm{R_c^2} + \frac{v^2\mathrm{b}^2t^2}{\mathrm{R_c^2}}\right)} \qquad [3.63c]$$

The expression [3.62] then allows $\mathrm{R_c}$ to be determined as a function of b, the interaction potential and the initial velocity $v = \sqrt{\frac{2\mathrm{E}}{\mu}}$.

For impact parameters that are lower than the closest distance of approach, the model is not realistic and when the impact parameter is zero, the velocity is zero at the closest point and the differential cross-section diverges.

Gersten [GER 71] and Bonamy *et al.* [BON 77] improved this model using a parabolic trajectory model and an equivalent rectilinear trajectory model to describe the relative trajectory of the collision partners.

The parabolic trajectory model is obtained from a limited expansion up to the second order in t of $\vec{R}(t)$ around $\vec{R}_c$. For expressions of $\vec{R}(t)$ and $\theta(t)$, one obtains:

$$R(t) = \sqrt{R_c^2 + (v_c' t)^2} \qquad [3.63d]$$

with: $v_c'^2 = v_c^2 + \vec{\gamma}_c \cdot \vec{R}_c$

and:

$$\theta(t) = \text{Arc sin}\left(\frac{v_c t}{R(t)}\right) = \text{Arc cos}\left(\frac{R_c + \frac{1}{2}\gamma_c t^2}{R(t)}\right) \qquad [3.63e]$$

It is thus better to include the real dependence of the modulus of R(t) as a function of time around the closest distance of approach. In the case where $v_c' = 0$, the collisions are orbiting because the centrifugal force and the attractive force offset each other. The duration of a collision is then greatly increased and the impact approximation is no longer justified because it assumes that the collisions are instantaneous.

The equivalent rectilinear trajectory model is defined with respect to $\vec{R}_c$ and constitutes an intermediate approximation of the real trajectory. The distance R(t) is processed up to the second order just like with the parabolic model. However, the angle $\theta(t)$ is approximated by:

$$\theta(t) = \text{Arc sin}\left(\frac{v_c' t}{R(t)}\right) = \text{Arc cos}\left(\frac{R_c}{R(t)}\right) \qquad [3.63f]$$

The two trajectory models can be used in Robert and Bonamy's theory.

3.2.9. *Expression of S(b,v) in terms of resonance functions*

The collision differential cross-section $S_2(\mathrm{b},v)$ is expressed as a function of the matrix elements of the collision operator P (equation [3.40]), meaning that:

$$\langle \mathrm{f}|\mathrm{P}|\mathrm{i}\rangle = \frac{i}{\hbar}\int_{-\infty}^{+\infty} \exp(i\omega_{\mathrm{if}}t)\langle \mathrm{f}|V(t)|\mathrm{i}\rangle\, dt \qquad [3.64]$$

where $|\mathrm{i}\rangle = |\mathrm{J}_1''\mathrm{M}_1''\mathrm{J}_2''\mathrm{M}_2''\rangle$ and $|\mathrm{f}\rangle = |\mathrm{J}_1'\mathrm{M}_1'\mathrm{J}_2'\mathrm{M}_2'\rangle$ are the eigenstates of the non-perturbed Hamiltonian H_0 of the partner molecules of a collision; ω_{if} is the angular frequency that corresponds to the transition between the states $|\mathrm{i}\rangle$ and $|\mathrm{f}\rangle$, given by: $\hbar\omega_{\mathrm{if}} = E_{\mathrm{J}_1'\mathrm{J}_2'} - E_{\mathrm{J}_1''\mathrm{J}_2''}$ and $V(t)$ is the molecular interaction potential.

For example, when one calculates the self-broadening parameter of a diatomic molecule in the general case, the purely electrostatic part of the interaction potential limited to the first four terms leads to the expression:

$$V_{\mathrm{E}}\left(\vec{\mathrm{R}}(t)\right) = V_{\mu_1\mu_2}\left(\vec{\mathrm{R}}(t)\right) + V_{\mu_1\mathrm{Q}_2}\left(\vec{\mathrm{R}}(t)\right) + V_{\mathrm{Q}_1\mu_2}\left(\vec{\mathrm{R}}(t)\right) + V_{\mathrm{Q}_1\mathrm{Q}_2}\left(\vec{\mathrm{R}}(t)\right) \qquad [3.65]$$

where $\vec{\mathrm{R}}(t)$ represents the intermolecular distance at the instant t, and μ_i and Q_i correspond to the dipole and quadrupole moments of the molecule i. When the molecule is homonuclear like O_2 or N_2, only the last term of equation [3.65] subsists. Thus, the differential cross-section is written as a sum of four terms (given the restriction concerning the homonuclear molecules) such as:

$$S_2(\mathrm{b},v) = S_2(\mathrm{b},v)_{\mu_1\mu_2} + S_2(\mathrm{b},v)_{\mu_1\mathrm{Q}_2} + S_2(\mathrm{b},v)_{\mathrm{Q}_1\mu_2} + S_2(\mathrm{b},v)_{\mathrm{Q}_1\mathrm{Q}_2} \qquad [3.66]$$

In this expression, each term corresponds to the different contributions of the electrostatic potential. When the contributions of the short- to mid-range interaction forces modeled by the atom–atom potential are taken into account, $S_2(\mathrm{b},v)$ is expressed as a sum of terms corresponding to the three types of the following terms, namely:

– the terms corresponding to the contribution of the electrostatic potential;

– the terms corresponding to the contribution of the atom–atom potential;

– the crossed terms simultaneously corresponding to the two types of contributions.

The time dependence of the interaction potential through the variables R(t) and $\theta(t)$ describing the trajectory of the collision partners, based on equation [3.64], leads to the evaluation of the integrals that are expressed in terms of the modified Bessel functions ($\mathrm{K}_m(\mathrm{k})$) [GRA 66] such as:

$$\int_{-\infty}^{+\infty}\frac{\mathrm{e}^{i\mathrm{kz}}P_l^m(\cos\theta)}{\left(1+\mathrm{z}^2\right)^{(l+1)/2}}dz=\frac{(-1)^{l+m}2i^{(l-m)}}{(1-m)!}\mathrm{k}^l\mathrm{K}_m(\mathrm{k}) \qquad [3.67]$$

where $P_l^m(\cos\theta)$ is a Legendre polynomial, $l=l_1+l_2$, z is a function of the collision parameters, k is a function of the collision parameters and the transition frequency ω_{if} and $\cos\theta=1/\sqrt{1+\mathrm{z}^2}$.

For the rectilinear trajectory, we have:

$$\mathrm{k}=\frac{\omega_{\text{if}}\mathrm{b}}{v},\quad \mathrm{z}=\frac{\mathrm{b}t}{v}\text{ and }\cos\theta=1/\sqrt{1+\mathrm{z}^2} \qquad [3.68a]$$

For the equivalent rectilinear trajectory, we have:

$$\mathrm{k}=\frac{\omega_{\text{if}}\mathrm{R_c}}{v'_\mathrm{c}},\quad \mathrm{z}=\frac{\mathrm{R_c}t}{v'_\mathrm{c}}\text{ and }\cos\theta=1/\sqrt{1+\mathrm{z}^2} \qquad [3.68b]$$

These integrals are generally expressed using the functions defined by [GRA 66]:

$$\mathrm{f}_l^m(\mathrm{k})=\frac{(-1)^l\,i^{(l-m)}}{2^{(2l-1)/2}}\left[\frac{(2l)!}{(l+m)!(l-m)!}\right]\mathrm{k}^l\mathrm{K}_m(\mathrm{k}) \qquad [3.69]$$

For an interaction of a given type, meaning a given l, we are led to perform a sum on the quantum numbers m and it appears in the differential cross-section's contribution, in the sum which is termed the resonance function and which is given by:

$$f_{l-1}(k) = \sum_m \left| f_l^m(k) \right|^2 \qquad [3.70]$$

with $f_{l-1}(0) = 1$.

For a quadrupole–quadrupole type interaction, for example, the resonance function corresponding to $S_2(b,v)_{outer,i}$ is expressed as:

$$f_3(k) = \frac{k^8}{2304}\left[K_4^2 + 8K_3^2 + 28K_2^2 + 56K_1^2 + 35K_0^2\right] \qquad [3.71]$$

When the trajectory is described by the parabolic model, we have:

$$k = \frac{\omega_{if} R_c}{v'_c}, \quad z = \frac{R_c t}{v'_c} \text{ and } \cos\theta = \left(1 - \left(v_c / v'_c\right)^2\right) z^2 \Big/ \left(\sqrt{1+z^2}\right) \qquad [3.72]$$

The form of $\cos\theta$ leads to the evaluation of the integrals whose expressions are given by:

$$\int_{-\infty}^{+\infty} \frac{e^{ikz} z^n}{\left(1+z^2\right)^{(l+1)/2}} dz \qquad [3.73]$$

In this case, the corresponding resonance functions depend not only on k but also on the $\frac{v_c}{v'_c}$ ratio [ROB 79]. The expressions of the different contributions to $S_2(b,v)$ of the resonance functions for a parabolic trajectory are given in [ROB 79]. Those corresponding to an equivalent rectilinear trajectory can be determined by replacing v_c with v'_c.

When one uses the parabolic trajectory model and the equivalent rectilinear trajectory model, the average on the impact parameter must be replaced by an average on the minimal approach distance. Based on the energy and angular momentum conservation laws, at the closest distance approach, it can be written that: $\frac{b}{R_c}\frac{db}{dR_c} = \left(\frac{v'_c}{v_c}\right)^2$ and then by replacing the average on the impact parameter by an average on the minimal approach distance, it follows:

$$\int_0^{+\infty} 2\pi \mathrm{b}\, d\mathrm{b} \rightarrow \int_{\mathrm{r}_0}^{+\infty} 2\pi\, \mathrm{R_c} \left(\frac{v_\mathrm{c}'}{v_\mathrm{c}}\right)^2 d\mathrm{R_c} \qquad [3.74]$$

where v is the average velocity and r_0 represents the minimal approach distance when the impact parameter b is zero.

3.3. True shape, profile and intensity of an absorption line

The intensity $I(\sigma)$ of a radiation of the wave number σ passing through an elementary thickness dl of a gas at a pressure P varies by a quantity $dI(\sigma)$ given by the Bouguer–Lambert law as:

$$dI(\sigma) = -k(\sigma,P)\, I(\sigma)\, dl \qquad [3.75]$$

where $k(\sigma,P)$ is the absorption coefficient with the wave number σ and with the pressure P. By integrating the previous equation, one extracts the intensity of the radiation transmitted at the wave number σ:

$$I_t(\sigma) = I_0(\sigma) e^{-k(\sigma,P)l} \qquad [3.76]$$

where $I_0(\sigma)$ is the intensity of the incident radiation on the gas and l is the thickness of the gas through which the radiation passes.

Furthermore, the true shape of an absorption line is defined by the function:

$$A(\sigma) = \frac{I_a(\sigma)}{I_0(\sigma)} = 1 - \frac{I_t(\sigma)}{I_0(\sigma)} \qquad [3.77]$$

disregarding the apparatus function of the measurement device, with $I_a(\sigma) = I_0(\sigma)\text{-}I_t(\sigma)$, in which $I_a(\sigma)$ is the intensity of the absorbed radiation.

Given the two previous equations, we have:

$$A(\sigma) = 1 - e^{-k(\sigma,P)l} \qquad [3.78]$$

The variation of the absorption coefficient $k(\sigma,P)$ with the wave number σ:

$$k(\sigma,P)=\frac{1}{l}\ln\left(\frac{I_0(\sigma)}{I_t(\sigma)}\right) \quad [3.79]$$

defines the profile of the spectral line.

When the number of active molecules (absorbing the radiation) is low, the product $k(\sigma,P)\ l$ verifies $k(\sigma,\ P)\ l \ll 1$. The true line shape function is written as:

$$A(\sigma)\approx k(\sigma,P)l \quad [3.80]$$

The profile of the line then tends toward its shape.

The quantity:

$$S=\int_0^\infty k(\sigma,P)d\sigma \quad [3.81]$$

leads to the line's integrated absorption coefficient.

If the active molecules absorb each other independently (de Beer hypothesis), this coefficient is proportional to the pressure, and it can be written that:

$$S=S^0P \quad [3.82]$$

in which S^0 is called "line intensity".

It should be noted that de Beer hypothesis is generally not verified for each wave number, the integrated absorption coefficient is then not proportional to the pressure for each wave number.

3.4. Line profile

The profile of a spectral line depends on the mechanisms responsible for its broadening.

Three factors participate in the width of a line and are divided into two categories:

– two broadening factors concern each of the active molecules, namely the "natural width" of its energy levels and the perturbations due to the other gas molecules. The line profile is then described by a Lorentz function. It is also called "homogeneous profile";

– in a very low-pressure gas, the effect due to the molecular interactions is negligible and the width of a line only depends on the Doppler–Fizeau effect connected to the relative movement of each of the molecules with respect to the observer. The line profile is described by a Gauss function. It is also called "heterogeneous profile".

In this case, the broadening by the pressure, assuming that the molecules are all moving at the average molecular velocity (by ignoring their Maxwellian distribution) and the Doppler–Fizeau effect, assuming that the molecules' trajectory is rectilinear (by neglecting the collisions) are considered to be statistically independent. The most general line profile is the convolution of the Lorentzian profile and the Gaussian profile, which results in the Voigt profile or Voigt function.

3.4.1. *Lorentz profile*

The profile of the line $K(\sigma,P)$ is given by the Lorentz function in the molecule's reference frame by:

$$K(\sigma,P)=\frac{S}{\pi}\frac{\gamma}{(\sigma-\sigma_0-\Delta)^2+\gamma^2} \qquad [3.83]$$

where S, σ_0, γ and Δ are the line's integrated absorption coefficient, the wave number at the line's center, its homogeneous width and its pressure shift, respectively.

In the absence of molecular interactions, at a very low pressure, we have:

$$\gamma=\gamma_N=\frac{1}{4\pi c t_{rad}} \qquad [3.84]$$

where γ_N is the line's natural width, t_{rad} is the radiative lifetime and c is the speed of light. When the effects of the pressure are no longer negligible, the width of the line also depends on the pressure and can be expressed by:

$$\gamma = \gamma_N + \gamma_L \approx \gamma_L \qquad [3.85]$$

where γ_L the half-width of the line profile is proportional to the frequency of the collisions and is very large compared to γ_N. For pressures which are not too high, it can be written that:

$$\gamma = \gamma^0 P \qquad [3.86]$$

where γ^0 is the broadening coefficient by the pressure P or broadening parameter, which is expressed in (10^{-3} cm^{-1} atm^{-1}) when the pressure is expressed in atmosphere. The integrated absorption coefficient is expressed as $S = \pi\gamma_L K_m$ in which K_m is the maximum absorption coefficient at the profile's center.

3.4.2. *Gauss profile*

The broadening parameters are measured in the laboratory reference frame and consequently, at non-zero temperature, the velocity of the molecule with respect to the observer must also be taken into account to determine the profile. The broadening by the Doppler–Fizeau effect requires the wave number σ to be known, in the laboratory reference frame, in which the molecule moving at the velocity $\vec{v}$ absorbs.

From the Lorentz transformation formula, we know that a molecule moving with a velocity $\vec{v}$, in the laboratory reference frame, absorbs a radiation at a wave number σ given by:

$$\sigma = \sigma_0 \left(1 + \frac{\vec{v} \cdot \vec{k}}{c} \right) \qquad [3.87]$$

where σ_0 is the wave number at the center of the line profile corresponding to the transition between the molecules' energy levels, $\vec{k}$ is a unitary vector in the direction of the beam and c is the speed of light.

The velocity distribution in the gaseous phase is given by the Maxwell–Boltzmann distribution law, such as:

$$dN_v = N\left(\frac{\mathrm{m}}{2\pi k_B T}\right)^{3/2} \exp\left(-\left(\frac{\mathrm{m}v^2}{2k_B T}\right)\right) d^3v \qquad [3.88]$$

where N is the number of gas molecules per unit volume, m is the mass of a molecule, v is the velocity of a molecule, T is the temperature, k_B is the Boltzmann constant and dN_v is the number of molecules per unit volume with the velocity ranging between v and $v+dv$.

From this, it results that the line profile broadened by the Doppler–Fizeau effect is a Gauss function, which is written as:

$$K_D(\sigma, P) = \sqrt{\frac{\mathrm{Log}2}{\pi}}\frac{S}{\gamma_D}\exp\left(-\mathrm{Log}2 \times \left(\frac{\sigma - \sigma_0}{\gamma_D}\right)^2\right) \qquad [3.89]$$

with γ_D the Doppler half-width equal to $\gamma_D = 3.58 \times 10^{-7} \sigma_0 \sqrt{\frac{T}{\mathrm{m}}}$, where m is the molecular mass of the absorbing molecule, T the temperature and σ_0 the wave number at the line center.

3.4.3. *Voigt profile*

When the two broadening mechanisms, the Doppler–Fizeau effect and the molecular interactions that are exerted throughout the collisions, contribute to the broadening of a line, the shape of the line is given by the convolution product of a Gaussian and a Lorentzian, such as:

$$K_V(\sigma, P) = K_D(\sigma_0, P)\frac{y}{\pi}\int_{-\infty}^{+\infty} \frac{\exp(-t^2)}{y^2 + (x-t)^2} dt \qquad [3.90]$$

with:

$$y=\frac{\gamma_L}{\gamma_D}\sqrt{\text{Log}2}\,,\, x=\frac{\sigma-\sigma_0-\Delta}{\gamma_D}\sqrt{\text{Log}2} \text{ and } K_D\left(\sigma_0,P\right)=\frac{S}{\gamma_D}\sqrt{\frac{\text{Log}2}{\pi}}$$

This function is called the Voigt function. For $y>>1$, the Voigt profile tends toward the Lorentz profile and for $y<<1$, the Voigt profile tends toward the Doppler–Fizeau profile. In some way, this is a representation of the "first order", when the free average path of the absorbers is much greater than the wave length of the radiation, i.e. in the pure Doppler regime.

3.4.4. *Galatry, Nelkin–Ghatak and Rautian–Sobelmann profiles*

Depending on the gas pressure, the narrowing effects can also be observed, generally because of the concomitance of two effects: the velocity dependence of the collision relaxation rate and the average effect of the velocity change in the collisions responsible for the broadening, which narrows the Doppler width. The collisional narrowing phenomena was first predicted by Dicke in 1953 [DIC 53] and observed for magnetic dipole transitions in microwaves.

Consequently, the "second-order" corrections, for an isolated line under the influence of binary collisions, should account for the appearance of a velocity dependence in the broadening by collision and the shift, as well as Dicke's narrowing effect. The collisional broadening velocity dependence causes the line to narrow, whereas the velocity dependence of the collisional shift leads to a line asymmetry. Dicke's narrowing takes place when the average free path of the absorbers subjected to the characteristic collisions by the velocity change are of the same order or lower than the wave length of the radiation.

The physical process of Dicke's narrowing can be modeled using the soft or hard collision approximation. In the first case, the effect of an individual collision is assumed to be infinitesimal, so that several collisions are necessary to produce a significant velocity variation. The absorber's movement becomes diffusive, leading to a Galatry profile [GAL 61] whose complex expression is given by:

$$G(x,y,z)=\sqrt{\frac{1}{\pi}}\mathrm{Re}\int_0^{+\infty}\exp\left(ixt-yt+\frac{1-zt-e^{-zt}}{2z^2}\right)dt \qquad [3.91]$$

with: $y=\frac{\gamma_L}{\gamma_D}\sqrt{\mathrm{Log}2}$, $x=\frac{\sigma-\sigma_0-\Delta}{\gamma_D}\sqrt{\mathrm{Log}2}$ and $z=\frac{\beta_L}{\gamma_D}\sqrt{\mathrm{Log}2}$, where β_L is the frequency of the collisions characterized by the velocity change.

The hard collision limit assumes that each collision completely affects the velocity, according to a Maxwellian distribution. This approximation leads to a Nelkin–Ghatak profile [NEL 64] or to a Rautian–Sobelmann profile [RAU 67].

$$R(x)=\sqrt{\frac{1}{\pi\gamma_L}}\mathrm{Re}\left(\frac{w(x,z)}{1-\sqrt{\pi}zw(x,z)}\right) \qquad [3.92]$$

where

$$w(x,y)=\frac{i}{\pi}\int_{-\infty}^{+\infty}\frac{\exp\left(-t^2\right)}{x-t+iy}dt \qquad [3.93]$$

is a complex probability function with:

$x=\frac{\sigma-\sigma_0-\Delta}{\gamma_D}\sqrt{\mathrm{Log}2}$, $y=\frac{\gamma_L}{\gamma_D}\sqrt{\mathrm{Log}2}$ and $z=\frac{\beta_L}{\gamma_D}\sqrt{\mathrm{Log}2}$, where β_L is the frequency of the collisions characterized by the velocity changes.

3.5. Conclusion

The profile of a spectral line depends on the mechanisms responsible for its broadening. Robert and Bonamy's theory, based on the impact approximation hypothesis (statistically independent and instantaneous binary collisions), on the notion of a classical trajectory (the dimension of the wave packet associated with a moving molecule is negligible with respect to the range of intermolecular interaction forces) and the non-overlapping of the spectral lines (isolated lines) can be applied to calculate line shifts and line broadenings in the infrared range. In the theory elaborated by Boulet and Robert [BOU 82], beyond this hypothesis, the overlapping effects of the lines can also be processed. This chapter mainly describes Robert and Bonamy's model, and in this chapter we have indicated how the works of other authors have improved its application. The interferences of lines have not been discussed, but the reader may find a more complete description of the effect of pressure on the shifts and widths of lines in Hartman *et al.*'s book [HAR 08].

3.6. Appendix

3.6.1. *Liouville formalism*

In the Liouville formalism, the Liouville space vectors are defined as tensor products of the vectors of a Hilbert space by the transposed conjugate vectors of this same space, such that:

$$\left|if\right\rangle\rangle = \left|i\right\rangle \times \left|f\right\rangle^{\dagger} \equiv \left|i\right\rangle \times \left\langle f\right| \qquad [3.94]$$

where $|i\rangle$ and $|f\rangle$ represent the initial and final states associated with the spectral line. This space comprises a "Liouvillian" operator, defined as:

$$L = HI - IH \qquad [3.95]$$

I being the unit operator, and whose action on any operator Θ of Hilbert space is defined by:

$$L\Theta = H\Theta - \Theta H \qquad [3.96]$$

These Liouville space properties are particularly useful for studying spectral lines (it is also called "line space"). The Liouville space vectors verify the orthogonality property:

$$\langle\langle if \left| i'f' \right\rangle\rangle = \left(\left|i\right\rangle\left\langle f\right|\right)^{\dagger} \times \left(\left|i'\right\rangle\left\langle f'\right|\right) = \left\langle i \middle| i' \right\rangle \times \left\langle f' \middle| f \right\rangle = \delta_{ii'}\delta_{ff'} \qquad [3.97]$$

where δ is the Kronecker symbol. Thus, the matrix element of the $L\Theta$ operator is written as:

$$\langle i|L\Theta|f\rangle = \sum_{i'f'} \langle\langle if|L|i'f'\rangle\rangle \langle i'|\Theta|f'\rangle \quad [3.98]$$

with, $\langle\langle if|L|i'f'\rangle\rangle = \langle\langle if|HI - IH|i'f'\rangle\rangle = \langle\langle if|HI|i'f'\rangle\rangle - \langle\langle if|IH|i'f'\rangle\rangle$, so that $\langle\langle if|L|i'f'\rangle\rangle = \langle i|H|i'\rangle \delta_{ff'} - \langle f'|H|f\rangle \delta_{ii'}$.

If $|i\rangle$ and $|f\rangle$ are eigenvectors of the Hamiltonian H, then we have:

$$\langle\langle if|L|i'f'\rangle\rangle = E_i \delta_{ii'} \delta_{ff'} - E_f \delta_{ii'} \delta_{ff'} = \hbar\omega_{if} \delta_{ii'} \delta_{ff'} \quad [3.99]$$

The Liouvillian L eigenvalues are therefore the energies of transitions and no longer the energies of levels as in the Hilbert space. Finally, the matrix element of the $L\Theta$ operator is written as:

$$\langle i|L\Theta|f\rangle = \sum_{i'} \left[\langle i|H|i'\rangle\langle i'|\Theta|f\rangle - \langle i|\Theta|i'\rangle\langle i'|H|f\rangle\right] \quad [3.100]$$

Furthermore, the time evolution equation of the operator Θ:

$$\dot{\Theta} = i\hbar^{-1}[H,\Theta] = i\hbar^{-1}L\Theta \quad [3.101]$$

has the following solution:

$$\Theta(t) = e^{i\hbar^{-1}Lt}\Theta(0) = e^{i\hbar^{-1}Ht}\Theta(0)e^{-i\hbar^{-1}Ht} \quad [3.102]$$

3.6.2. *The Clebsch–Gordan coefficients and the Wigner 3j symbols*

The coupled states of the spherical harmonics can be expanded using the closure relation in the decoupled basis such as:

$$|JM\rangle = \sum_{m_1=-j_1}^{m_1=+j_1} \sum_{m_2=-j_2}^{m_2=-j_2} |j_1m_1j_2m_2\rangle\langle j_1m_1j_2m_2|JM\rangle \quad [3.103]$$

where $\langle j_1 m_1 j_2 m_2 | JM \rangle$ are the Clebsch–Gordan coefficients. They are sometimes written in the following form $\langle j_1 j_2; m_1 m_2 | JM \rangle$ or $C(j_1 j_2 J; m_1 m_2 M)$. The quantum numbers j_1, j_2 and J verify the condition: $j_1 - j_2 \leq J \leq j_1 + j_2$ and the magnetic quantum numbers m_1, m_2 and M verify the relation: $m_1 + m_2 = M$.

The Clebsch–Gordan coefficients can be expressed according to the Wigner 3j symbols, which are easier to use given their symmetry property, as follows:

$$\begin{pmatrix} j_1 & j_2 & J \\ m_1 & m_2 & -M \end{pmatrix} = \frac{(-1)^{j_1 - j_2 + M}}{\sqrt{2J+1}} \langle j_1 m_1 j_2 m_2 | JM \rangle \quad [3.104]$$

where the exponent j_1-j_2 + M is an integer. The inverse relation is expressed as:

$$\langle j_1 m_1 j_2 m_2 | JM \rangle = (-1)^{j_2 - j_1 - M} \sqrt{2J+1} \begin{pmatrix} j_1 & j_2 & J \\ m_1 & m_2 & -M \end{pmatrix} \quad [3.105]$$

The sum of the arguments $j_1 + j_2 + J$ is an integer.

3.6.3. *The terms of the differential cross-section expansion S(b,v)*

The expressions of the different terms that appear in the differential cross-section expansion $S(\mathrm{b},v)$ are as follows:

– Zeroth-order term:

$$S(\mathrm{b},v) = 0 \quad [3.106]$$

– First-order term:

$$S_1(\mathrm{b},v) = S_{1,\mathrm{i}}(\mathrm{b},v) - S_{1,\mathrm{f}}(\mathrm{b},v) \quad [3.107]$$

with

$$S_{1,i}(b,v) = \frac{i}{\hbar}\sum_{m_i m_2} \frac{\langle v_i j_i m_i v_2 j_2 m_2 | P | v_i j_i m_i v_2 j_2 m_2 \rangle}{2J_i + 1} \quad [3.108]$$

and

$$S_{1,f}(b,v) = \frac{i}{\hbar}\sum_{m_i m_2} \frac{\langle v_f j_f m_f v_2 j_2 m_2 | P | v_f j_f m_f v_2 j_2 m_2 \rangle}{2J_f + 1} \quad [3.109]$$

– Second-order term:

$$S_2(b,v)_{\text{outer},i} = \frac{1}{2}\sum_{m_i m_2} \frac{\langle v_i j_i m_i v_2 j_2 m_2 | P^2 | v_i j_i m_i v_2 j_2 m_2 \rangle}{2J_i + 1} \quad [3.110]$$

$$S_2(b,v)_{\text{middle}} = \frac{1}{2}\sum_{\substack{m_i m_f \\ m_i' m_f' \\ m}} \sum_{\substack{v_2' J_2' m_2' \\ m_2}} (-1)^{m_f - m_f'} \begin{pmatrix} J_f & 1 & J_i \\ -m_f & m & m_i \end{pmatrix} \begin{pmatrix} J_f & 1 & J_i \\ -m_f' & m & m_i' \end{pmatrix} \quad [3.111]$$
$$\langle v_f j_f m_f' v_2' j_2' m_2' | P | v_f j_f m_f v_2 j_2 m_2 \rangle \langle v_i j_i m_i v_2 j_2 m_2 | P | v_i j_i m_i' v_2' j_2' m_2' \rangle$$

$$S_2(b,v)_{\text{outer},f} = \frac{1}{2}\sum_{m_i m_2} \frac{\langle v_f j_f m_f v_2 j_2 m_2 | P^2 | v_f j_f m_f v_2 j_2 m_2 \rangle}{2J_f + 1} \quad [3.112]$$

4

Energy Levels and Spectral Profile of a Diatomic Molecule in Condensed Phase

In this chapter, we describe the theoretical models developed to analyze the absorption spectrum of a molecule, isolated in an inert matrix of rare gas or nitrogen at a very low temperature in a nanocage. The inclusion model leads to determining the trapping site (or sites) and the position as well as the movements of the molecule in its site. The frequency shifts due to the solid environment may be interpreted by using an atom–atom potential to describe the interaction between the matrix and the trapped molecule. The anisotropic distortion of the nanocage corresponding to the substitutional site is evaluated by an iterative method based on the Green's functions of the perfect crystal, the Green's matrix being the inverse of the force constant matrix. The line broadening as a function of the temperature is connected to the molecule–matrix coupling and, in some cases, may be described by a single law as deduced from the theories developed to interpret the experimental results on the couplings between a molecule and a solid or liquid environment (direct multiphononic transfer; binary collision). When the molecule is trapped in clathrate, fullerene or zeolite nanocages, we use the extended inclusion model to determine the IR spectra and the couplings with the atoms or the molecules constituting the nanocage.

4.1. Introduction

The construction of the spectral profile of a molecule trapped in a condensed environment, that is, the shape and the width of the spectral lines, was carried out when it was possible to compare the theoretical results with the high-resolution spectroscopic measurements.

The behavior of the molecules trapped in a matrix was analyzed for diatomic molecules such as the hydracids and the carbon monoxide trapped in simple matrices such as rare gas solids. These studies were focused on the quality of the molecule–matrix interaction potential, the symmetry of the inclusion site and the quality of the approximations in the decoupling of the movements by reconstituting the bar spectrum (frequencies and intensities of the lines) and by comparing it with the experimental spectra.

The first models proposed for interpreting the experimental spectra were developed by Pauling [PAU 30] and Devonshire [DEV 36]. The first one is based on the hindered (blocked) rotation of a diatomic molecule in the crystalline inclusion site. The second one consists of the study of the quasi-free rotation movement of the molecule whose center of mass is fixed at the substitutional site center of the static crystal with octahedral symmetry. These models were later refined and generalized by Flygare [FLY 63] for other types of site symmetry.

These models have enabled us to calculate frequency shifts in agreement with the experimental results. However, they were too simplistic to interpret the infrared spectra of molecules such as HCl and CO trapped in rare gas matrices.

To improve the description of the movements, Friedmann and Kimel [FRI 65] on the one hand, and Baur and Salzman [BAU 66] on the other hand, proposed a model called the "RTC" (rotation–translation coupling) by considering the coupling between the orientation of the diatomic molecule and the translation of its center of mass, the crystalline environment being considered static and undistorted (cage model). The basic hypothesis is that the center of interaction onto which the crystalline field is applied is different from the molecule's center of mass. Although this model has enabled certain spectral characteristics to be interpreted in the case of HCl trapped in an argon matrix, it could not be generalized. Indeed, the decoupling between the molecule's center of mass translational vibration mode with frequency ω_T and the vibration modes (phonons) of the crystalline environment is only possible if the frequency ω_T is much larger than the cut-off frequency of the perfect crystal's band of phonons. Taking this coupling into account, Mannheim and Friedmann [MAN 72] were able to reconstitute the spectral profile of the HCl and HBr molecules trapped in the argon matrix. However, the results were not satisfactory for other matrices and other diatomic molecules, such as CO for example.

The first model that accounts for the coupling between the movements of the diatomic molecule and those of the undistorted crystalline lattice (phonons) with octahedral symmetry was proposed by Kono and Lin [KON 83] for HCl and DCl trapped in rare gas matrices. In the first instance, the Born–Oppenheimer approximation is used to define the high- and low-frequency movements separately. The coupling between these movements is then introduced to calculate the frequency shifts and the line widths in order to reconstitute the spectral profile. The form of the interaction potential energy between the trapped molecule and the atoms of the crystal limited the model to the systems in which the size of the molecule is of the same order of magnitude as that of the trapping site.

In the Lakhlifi–Dahoo extended substitution model, the effective charges calculated by the quantum chemistry methods are used to determine the electrical field of the environment in which the molecule moves inside a nanocage. In this case, the effective charges are placed on each atom of the trapped molecule and of the molecules constituting the nanocage structure. In the case of rare gas matrices, the electrical field is due to the polarizabilities of the atoms constituting the matrix. The polarizability field is induced by the electrical moments of the trapped molecule.

4.2. Inclusion model

Constructing a theoretical model that allows the spectroscopic phenomena to be quantitatively interpreted, requires a good and accurate knowledge of the interaction potential energy between the trapped molecule and the atoms or molecules constituting the nanocage. This energy must also have an analytical form, which is explicitly dependent on the active molecule's degrees of freedom and its partners, namely their positions, orientation and internal vibrations.

The interaction energy is generally determined from pairwise interaction (binary interaction) between the trapped molecule and each atom or molecule of the nanocage. In fact, the environmental effects of each species are most of the time accounted for through effective parameters of this binary potential energy.

This binary potential energy is the sum of long-range contributions of electrostatic, inductive and dispersive nature (Van der Waals interactions) and short-range contributions of a repulsive nature, which describes, on the one hand, the overlapping of the electronic clouds of the two partners and, on the other hand, the possible charge transfer phenomena.

The analytical expressions of the electrostatic, inductive and dispersive parts of the energy are now well known for small systems through different molecular electrical characteristics, namely the multipole moments (dipole, quadrupole, octupole, etc.) and the electrical polarizabilities (dipolar, quadripolar, etc.). In contrast, regarding the repulsive interactions, there is no satisfactory description by analytical methods, and only quantum chemistry numerical techniques (semi-empirical, ab initio, etc.) can lead to a good description.

In section 4.2.1, we will briefly introduce the different energy contributions for a pair of partners: "optically active diatomic molecule–molecule or atom of the nanocage".

4.2.1. *Binary interaction energy*

Let us consider a diatomic molecule AB (noted as 0) interacting with a molecule of the nanocage (noted as k). The position and the orientation of each of these molecules are defined with respect to a fixed frame, which will later be associated with the nanocage. The molecules are assumed to be in their fundamental electronic state.

The molecules possess a charge distribution described in terms of a set of permanent multipole moments (dipole, quadrupole, etc.). This charge distribution may, under the influence of an electrical field, also undergo a non-permanent distortion. The molecules are then said to be "polarizable" and to be characterized by a polarizability (dipolar, quadrupolar, etc.). This distortion is modeled by introducing the concept of induced multipole moments.

Figure 4.1 shows the geometrical characteristics of the system formed by the two partner molecules, in which the molecule k is a molecule belonging to the nanocage. This may be an atom of a rare gas matrix, a N_2 molecule of a nitrogen matrix or even a H_2O molecule of a clathrate hydrate nanocage, for example.

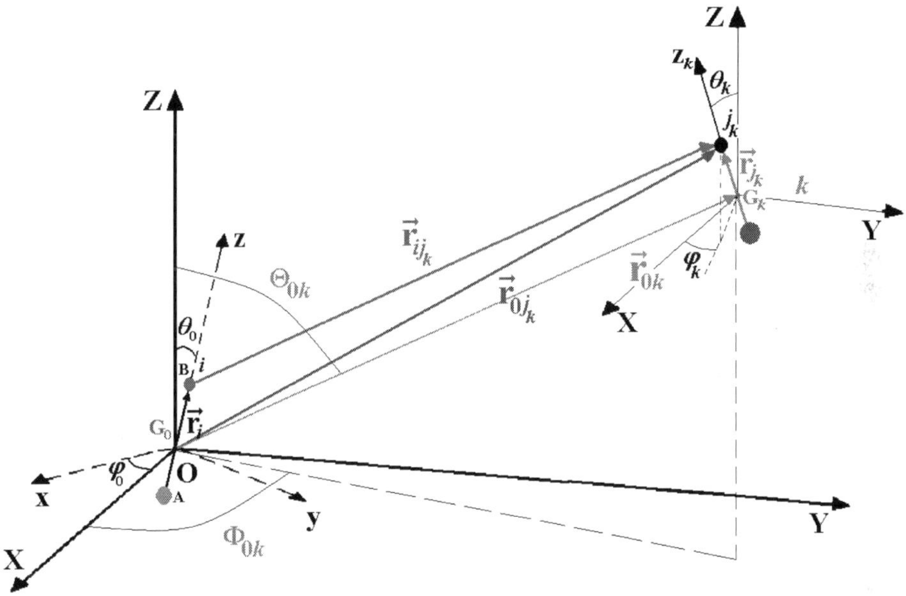

Figure 4.1. *Geometrical characteristics of a diatomic molecule* AB *interacting with a molecule of a nanocage (atom, diatomic or polyatomic). ($G_0,\vec{x},\vec{y},\vec{z}$) and (O,X,Y,Z), respectively, represent the frame tied to the studied molecule and the fixed frame (of the laboratory), which will be tied to the nanocage. For a color version of this figure, see www.iste.co.uk/dahoo/infrared.zip*

In this figure:

– $\vec{r}_{ij_k}$ and $\vec{r}_{0j_k}$ are, respectively, the distance vectors between the atom i of the diatomic molecule AB, and on the one hand, its center of mass, and on the other hand, the atom j of the molecule k of the nanocage;

– $\vec{r}_{0k}$ is the distance vector between the AB molecule's center of mass and that of the molecule k;

– $\vec{\mathrm{r}}_i$ and $\vec{\mathrm{r}}_{j_k}$ are the position vectors of the atom i of the molecule AB and the atom j of the molecule k with respect to their respective tied frame;

– (θ_0, φ_0) and (θ_k, φ_k) are the Euler angles characterizing the orientation of each molecule with respect to the fixed frame associated with its center of mass;

– finally, (Θ_{0k}, Φ_{0k}) are the Euler angles associated with the distance vector $\vec{\mathrm{r}}_{0k}$ with respect to the fixed frame.

The different contributions to the pairwise interaction energy "studied molecule – molecule of the nanocage" are split into three groups as described hereafter.

4.2.1.1. *Dispersion–repulsion contribution*

This part is generally modeled by a 12-6 Lennard–Jones (LJ) type atom–atom potential energy in the form:

$$V_{\mathrm{LJ}} = \sum_i \sum_{j_k} 4\varepsilon_{ij} \left\{ \left(\frac{\sigma_{ij}}{\left| \vec{\mathrm{r}}_{ij_k} \right|} \right)^{12} - \left(\frac{\sigma_{ij}}{\left| \vec{\mathrm{r}}_{ij_k} \right|} \right)^{6} \right\} \qquad [4.1]$$

On the one hand, the first term of this expression (in $1/\mathrm{r}^{12}$) is the short-range repulsion term, which is connected to the Pauli principle and the repulsion of the two interpenetrating electronic clouds of the two atoms. The second term (in $1/\mathrm{r}^{6}$) is the long-range attractive term, which is due to the interaction between induced multipoles on the two atoms.

On the other hand, the parameters of the LJ potential ε_{ij} and σ_{ij} are defined by the equations $\varepsilon = -V_{\mathrm{LJ}}^{\min}$ and $V_{\mathrm{LJ}}(\sigma) = 0$ (Figure 4.2). They are determined by the Lorentz–Berthelot combination rules:

$$\begin{aligned} \varepsilon_{ij} &= \sqrt{\varepsilon_{ii}\varepsilon_{jj}} \\ 2\sigma_{ij} &= \sigma_{ii} + \sigma_{jj} \end{aligned} \qquad [4.2]$$

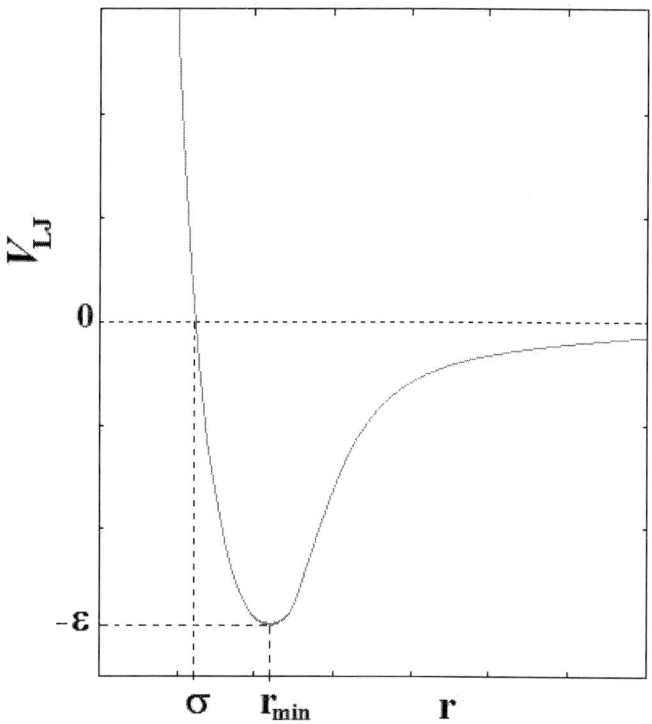

Figure 4.2. *Form of atom–atom interaction potential energy of the Lennard–Jones type*

4.2.1.2. *Electrostatic contribution*

This contribution is due to the interaction between the permanent multipole moments of the two molecules, which are assumed to be electrically neutral. However, depending on the distance separating the two molecules, these may interact through their charge distributions or through their permanent multipoles moments.

In fact, if the distance is of the same order of magnitude as the size of the molecules, the electrostatic energy is written as:

$$V_{\mathrm{E}} = \frac{1}{4\pi\varepsilon_0}\sum_i \sum_{j_k} \frac{q_i q_{j_k}}{\left|\vec{\mathrm{r}}_{ij_k}\right|} \qquad [4.3]$$

where q_i and q_{j_k} are, respectively, the charge of the site i of the molecule 0 and the charge of the site j of the molecule k, and $\vec{r}_{ij_k}$ is the distance vector between these two sites. We recall that, depending on the molecule, the sites carrying the charges may be different from the atoms centers constituting this molecule.

However, when the distance separating the two molecules is quite large compared to their size, each one seems rather neutral for the other one and the interaction occurs through their multipole moments.

Each molecule creates in its surroundings an electrical potential. These are the terms $\Phi_{0\to k}$ and $\Phi_{k\to 0}$ the electrical potentials created, respectively, by the molecule 0 at the point of position vector $\vec{r}_k$ (which is identical to $\vec{r}_{0k}$) and by the molecule k at point O (Figure 4.1), and which are defined by [BUC 67] :

$$\left\{\begin{array}{l} \Phi_{0\to k} = \dfrac{1}{4\pi\varepsilon_0}\left\{\begin{array}{l} -\vec{\nabla}\left(\dfrac{1}{\left|\vec{r}_{0k}\right|}\right).\vec{\mu}_0 + \dfrac{1}{3}\vec{\nabla}\vec{\nabla}\left(\dfrac{1}{\left|\vec{r}_{0k}\right|}\right):\vec{\vec{\Theta}}_0 \\ \qquad -\dfrac{1}{15}\vec{\nabla}\vec{\nabla}\vec{\nabla}\left(\dfrac{1}{\left|\vec{r}_{0k}\right|}\right)\vdots\vec{\vec{\vec{\Omega}}}_0 + \ldots \end{array}\right\} \\ \Phi_{k\to 0} = \dfrac{1}{4\pi\varepsilon_0}\left\{\begin{array}{l} -\vec{\nabla}\left(\dfrac{1}{\left|\vec{r}_{0k}\right|}\right).\vec{\mu}_k + \dfrac{1}{3}\vec{\nabla}\vec{\nabla}\left(\dfrac{1}{\left|\vec{r}_{0k}\right|}\right):\vec{\vec{\Theta}}_k \\ \qquad -\dfrac{1}{15}\vec{\nabla}\vec{\nabla}\vec{\nabla}\left(\dfrac{1}{\left|\vec{r}_{0k}\right|}\right)\vdots\vec{\vec{\vec{\Omega}}}_k + \ldots \end{array}\right\} \end{array}\right. \qquad [4.4]$$

where $\vec{r}_{0k}$ is the distance vector between the two molecules; $\vec{\mu}$, $\vec{\vec{\Theta}}$, $\vec{\vec{\vec{\Omega}}}$, etc. are their dipole (vector), quadrupole (rank 2 tensor), octupole (rank 3 tensor), etc. moments and ε_0 is the vacuum electrical permittivity. The symbol $\vec{\nabla}$ designates the gradient operator that transforms a function $f(X,Y,Z)$ into the vector:

$$\vec{\nabla} f = \begin{vmatrix} \dfrac{\partial f}{\partial X} \\ \dfrac{\partial f}{\partial Y} \\ \dfrac{\partial f}{\partial Z} \end{vmatrix}$$

The corresponding electrical fields are given by:

$$\begin{aligned} \vec{E}_{\mathbf{0}\to k} &= -\vec{\nabla}\Phi_{0\to k} \\ \vec{E}_{k\to 0} &= -\vec{\nabla}\Phi_{k\to 0} \end{aligned} \qquad [4.5]$$

The electrostatic interaction potential energy between the two molecules is then written as:

$$\begin{aligned} V_{\mathrm{E}} &= \vec{\mu}_0.\vec{\nabla}\Phi_{k\to 0} + \frac{1}{3}\vec{\vec{\Theta}}_0 : \vec{\nabla}\vec{\nabla}\Phi_{k\to 0} + \frac{1}{15}\vec{\vec{\vec{\Omega}}}_0 \vdots \vec{\nabla}\vec{\nabla}\vec{\nabla}\Phi_{k\to 0} + \ldots \\ &= \vec{\mu}_k.\vec{\nabla}\Phi_{0\to k} + \frac{1}{3}\vec{\vec{\Theta}}_k : \vec{\nabla}\vec{\nabla}\Phi_{0\to k} + \frac{1}{15}\vec{\vec{\vec{\Omega}}}_k \vdots \vec{\nabla}\vec{\nabla}\vec{\nabla}\Phi_{0\to k} + \ldots \end{aligned} \qquad [4.6]$$

Finally, by introducing equations [4.4] and [4.5] into one of equations [4.6] and by limiting the expression to the fourth order, we obtain:

$$V_{\mathrm{E}} = \frac{1}{4\pi\varepsilon_0}\left\{ \begin{aligned} &-\vec{\mu}_0.\vec{\nabla}\vec{\nabla}\left(\frac{1}{\left|\vec{r}_{\mathbf{0}k}\right|}\right).\vec{\mu}_k - \frac{1}{3}\vec{\mu}_0.\vec{\nabla}\vec{\nabla}\vec{\nabla}\left(\frac{1}{\left|\vec{r}_{\mathbf{0}k}\right|}\right) : \vec{\vec{\Theta}}_k \\ &+\frac{1}{3}\vec{\vec{\Theta}}_0 : \vec{\nabla}\vec{\nabla}\vec{\nabla}\left(\frac{1}{\left|\vec{r}_{\mathbf{0}k}\right|}\right).\vec{\mu}_k - \frac{1}{15}\vec{\mu}_0.\vec{\nabla}\vec{\nabla}\vec{\nabla}\vec{\nabla}\left(\frac{1}{\left|\vec{r}_{\mathbf{0}k}\right|}\right) \vdots \vec{\vec{\vec{\Omega}}}_k \\ &-\frac{1}{15}\vec{\vec{\vec{\Omega}}}_0 \vdots \vec{\nabla}\vec{\nabla}\vec{\nabla}\vec{\nabla}\left(\frac{1}{\left|\vec{r}_{\mathbf{0}k}\right|}\right).\vec{\mu}_k + \frac{1}{9}\vec{\vec{\Theta}}_0 : \vec{\nabla}\vec{\nabla}\vec{\nabla}\vec{\nabla}\left(\frac{1}{\left|\vec{r}_{\mathbf{0}k}\right|}\right) : \vec{\vec{\Theta}}_k + \ldots \end{aligned} \right\} \qquad [4.7]$$

4.2.1.3. *Induction contribution*

As said previously, with the molecules being polarizable, each one creates induced multipole moments on the other. The induction potential energy is due to the interaction between the induced moments and the permanent moments. It is written as:

$$V_{\mathrm{I}} = -\frac{1}{2}\vec{E}_{\mathbf{0}\to k}.\vec{\vec{\alpha}}_k.\vec{E}_{\mathbf{0}\to k} - \frac{1}{2}\vec{E}_{k\to 0}.\vec{\vec{\alpha}}_0.\vec{E}_{k\to 0} \qquad [4.8]$$

where $\vec{E}$ are the previously defined electrical fields (equations [4.5]) and $\vec{\vec{\alpha}}$ are the dipolar polarizabilities (rank 2 tensors) of the partner molecules.

As indicated previously, the interaction energy between the trapped molecule and the atoms or molecules of the nanocage must have an analytical form, which is explicitly dependent on the degrees of freedom of the molecule and its partners, namely their translation, orientation and internal vibration movements. This property is essential to process the coupling between the system's degrees of freedom.

In fact, very often this ends up determining the successive derivatives of the molecular parameters. So, an *exact* analytical form cannot be established without recourse to approximations and truncations in the mathematical developments so as to ensure a certain convergence.

It is therefore necessary to match the analytical developments and numerical calculations.

A two-center expansion [HIR 64] is used to make the term $1/\left|\vec{\mathrm{r}}_{ij_k}\right|$ explicit as a function of the different variables of the system:

$$\frac{1}{\left|\vec{\mathrm{r}}_{ij_k}\right|}=\frac{1}{\left|\vec{\mathrm{r}}_{0k}\right|}\left\{1+\sum_{\substack{l_0,l_k\\p_0,p_k}}^{*}\left[F_{p_0,p_k}^{l_0,l_k}(r_i,\beta_i,r_{j_k},\beta_{j_k})Y_{l_0}^{-p_0}(\theta_0,\phi_0)Y_{l_k}^{p_k}(\theta_k,\phi_k)\right]\right\} \quad [4.9]$$

with

$$\begin{aligned}F_{p_0,p_k}^{l_0,l_k}(r_i,\beta_i,r_{j_k},\beta_{j_k})=&\frac{(4\pi)^2(-1)^{l_k+p_k+p_0}}{(2l_0+1)(2l_k+1)}\left(\frac{(r_i)^{l_0}(r_{j_k})^{l_k}}{\left|\vec{\mathrm{r}}_{0k}\right|^{l_0+l_k}}\right)Y_{l_0}^{0}(\beta_i,0)\times\\&Y_{l_k}^{0}(\beta_{j_k},0)\sum_{m=-l_k}^{+l_k}\left\{\begin{array}{c}\dfrac{(l_0+l_k)!}{\left[(l_0+|m|)!(l_0-|m|)!(l_k+|m|)!(l_k-|m|)!\right]^{1/2}}\times\\ \displaystyle\sum_{l=|l_0-l_k|}^{l_0+l_k}\sqrt{\frac{4\pi}{(2l+1)}}C(l_0l_kl;m,-m)C(l_0l_kl;p_0,-p_k)\\ Y_l^{p_0-p_k}(\Theta_{0k},\Phi_{0k})\end{array}\right\}\end{aligned} \quad [4.10]$$

In equation [4.9]: l_0 and l_k are natural integers; p_0 and p_k are relative integers such as: $-l_0 \leq p_0 \leq l_0$ and $-l_k \leq p_k \leq l_k$; the sign (*) on the sum sign indicates that $l_0 + l_k \neq 0$, β_i and β_{j_k} are, respectively, the internal angular parameters of the atom *i* of the molecule 0 and of the atom *j* of the molecule *k* in their respective frame (for diatomic molecules, the only values taken by these angles are: 0 and π), and *Y* designate the spherical harmonics. In equation [4.10], $l_{\langle} = \inf\{l_0, l_k\}$ is the smaller of the two values and *C* are the Clebsch–Gordan coefficients.

Note that the dispersion–repulsion contribution of the potential energy depends on $\left(1/\left|\vec{r}_{ij_k}\right|\right)^6$ and $\left(1/\left|\vec{r}_{ij_k}\right|\right)^{12}$, which leads us to perform binomial expansions of expression [4.9]. The ratio $\dfrac{(r_i)^{l_0}(r_{j_k})^{l_k}}{\left|\vec{r}_{0k}\right|^{l_0+l_k}}$ that appears in equation [4.10] is the main term, which is essential in the convergence and the truncation of the series and binomial expansions.

Finally, in this chapter we will consider the inclusion of the diatomic molecule AB in a rare gas matrix, and as the molecule *k* is in fact an atom, we then use a one-center development. As an example, section 4.8.1 shows the expression obtained, in the fourth order of truncation expansion, of the dispersion–repulsion contribution.

4.2.2. *Lakhlifi–Dahoo inclusion model*

In the Lakhlifi–Dahoo extended approach, the long-distance electrostatic interactions between the trapped molecules and the host molecules are determined from the effective charges, either determined from the quantum chemistry *ab-initio* calculations, or adjusted to the experimental results on the shift or broadening of the lines. Thus, the static electric field present in the nanocage is determined from effective charges placed on the molecules, which make up the nanocage. The perturbation of the electronic potential driving the nuclear movements of the trapped molecule is then calculated from the interaction between these effective charges and those of the trapped molecule. A shielding effect that accounts for the non-spherical character of the interaction is applied by adjusting the available experimental values on the shifts of the lines from an isotopologue to those calculated in the corresponding type of nanocage. The same shielding effect is then considered to calculate the infrared spectra of the isotopologues in the different structures or nanocages. For example, when the molecule is trapped in a clathrate cage, the interaction at the short range with water molecules modifies the translation and rotation degrees of freedom as the molecule is no longer free to move. Consequently, only the Q branches are present in the molecule's infrared spectra.

4.3. Rare gas nanocage

4.3.1. *The rare gases in the solid state*

The solid rare gases (matrices) are inert crystals composed of weakly bound atoms. Their cohesion is ensured by Van der Waals type electrostatic bonds so that they crystallize at very low temperatures. Their cohesion energy is weak (Table 4.1).

Rare gas	Neon	Argon	Krypton	Xenon
Fusion temperature (K)	25	84	115	161
Cohesion energy (kcal/mole)	0.45	1.85	2.67	3.83

Table 4.1. *Temperature and cohesion energy of the rare gas matrices*

In rare gas matrices, the interatomic forces are generally assumed to be central and binary. Various interaction potential energy models have been proposed and among these, the LJ (12-6) model, defined in equation [4.1], and that of Buckingham (6-exponential) are largely used. The difference between these two models resides in the expression of the repulsive part, in $1/r^{12}$ for the first one, and in e^{-ar} for the second. In this work, only the model (LJ) is used.

The interaction potential energy in a rare gas perfect crystal can therefore be written as:

$$V_C = \sum_j \sum_{j'>j} 4\varepsilon_{\text{gr-gr}} \left\{ \left(\frac{\sigma_{\text{gr-gr}}}{\left|\vec{r}_{jj'}\right|} \right)^{12} - \left(\frac{\sigma_{\text{gr-gr}}}{\left|\vec{r}_{jj'}\right|} \right)^{6} \right\} \quad [4.11]$$

The pure parameters $\varepsilon_{\text{gr-gr}}$ and $\sigma_{\text{gr-gr}}$ are given in Table 4.2. We note that in Figure 4.1, the molecule *k* of the nanocage is replaced by the atom *j*. The *k* index is then useless.

Rare gas	Ne	Ar	Kr	Xe
$\varepsilon_{\text{gr-gr}}$ (cm^{-1})	30.0	84.0	125.0	157.1
$\sigma_{\text{gr-gr}}$ (Å)	2.764	3.448	3.650	3.970

Table 4.2. *Parameters of the Lennard–Jones potential of rare gases*

The rare gases generally crystallize in a face-centered cubic system (fcc). However, when impurities are present, they can be stabilized in the hexagonal compact structure (hcp). Moreover, the rigidity of the matrix increases with the size of the rare gas atoms (Ne–Ar–Kr–Xe).

Different trapping sites are possible in rare gas matrices. Depending on the size and shape of the trapped molecule, it may substitute one, two or more atoms of the lattice or, if it is small enough, it may be inserted into an interstitial site.

The fcc structure of the rare gas matrices belongs to the octahedral symmetry group O_h. The unit cell is a cube of edge *a* (cell parameter) and includes eight nodes at the corners and six nodes at the center of the faces. The primitive cell is formed by the primitive vectors $\vec{a}_1$, $\vec{a}_2$ and $\vec{a}_3$, which connect a node located at the corner of the cube to those located in the center of the faces as shown in Figure 4.3. Thus, an atom *j* of the lattice is labeled by its equilibrium position vector $\vec{R}_j^{\text{e}}$ as:

$$\vec{R}_j^{\text{e}} = n_1(j)\vec{a}_1 + n_2(j)\,\vec{a}_2 + n_3(j)\vec{a}_3 \quad [4.12]$$

Each atom is then characterized by the set of three numbers (n_1, n_2, n_3).

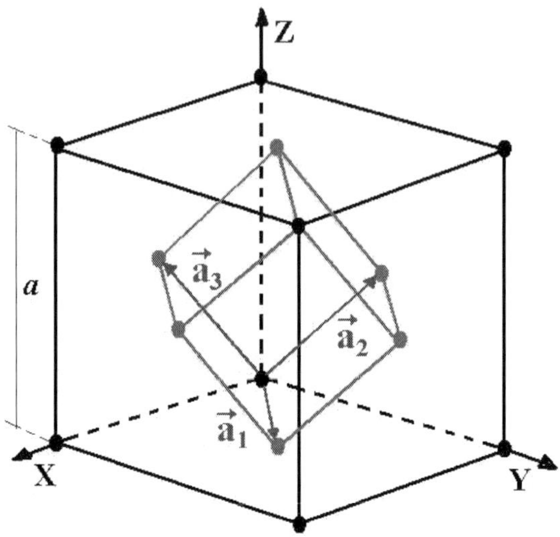

Figure 4.3. *Elementary cell of a face-centered cubic crystal*

Conventionally, the crystallographic characteristics of the fcc lattice are:

– number of nodes per cell: 4;

– volume of the primitive cell: a^3;

– number of nearest neighbors: 12;

– distance between the nearest neighbors: $a/\sqrt{2}$;

– number of second neighbors: 6;

– distance between second neighbors: a;

– number of third neighbors: 24;

– distance between third neighbors: $a\sqrt{2}$.

Matrix	Ne	Ar	Kr	Xe
a (Å)	4.34	5.30	5.67	6.10

Table 4.3. *Cell parameter in the rare gas matrices*

4.3.2. *Dynamics of the perfect fcc lattice (Bravais lattice)*

In order to determine the density of the vibration modes of the crystal lattice or phonon density, one must solve the eigenvalues equation associated with the Hamiltonian of this lattice.

In the harmonic approximation, the dynamic Hamiltonian of a three-dimensional (3N dimensional) monoatomic (N atoms) crystal lattice is given by [HER 59, SCH 52]:

$$H_C = T_C + V_C = \sum_{\alpha,j} \frac{P_\alpha^2(j)}{2M_j} + \frac{1}{2} \sum_{\alpha,j} \sum_{\beta,j'} A_{\alpha\beta}(j,j')u_\alpha(j)u_\beta(j') + V_C^e \qquad [4.13]$$

where T_C is the translation kinetic Hamiltonian of the crystal atoms, V_C its potential energy and V_C^e its potential energy at equilibrium (rigid crystal), and:

– $u_\alpha(j)$ and $u_\beta(j')$ are the translation dynamic variables of the atoms j and j' following the directions α and β (X, Y, Z), around their equilibrium position, with respect to the frame associated with the crystal (absolute frame);

– $P_\alpha(j)$ and M_j are, respectively, the component following the direction α of the conjugate momentum of the dynamic translation variable $u_\alpha(j)$ of the atom j and its mass. In fact, as the lattice is perfect, the N atoms have the same mass;

– and $A_{\alpha\beta}(j,j')$ represents the second-order force constant, meaning the force exerted along the direction α on the atom j when the atom j' moves along the direction β. This is the second derivative of the potential energy:

$$A_{\alpha\beta}(j,j') = \left(\frac{\partial^2 V_C}{\partial u_\alpha(j) \partial u_\beta(j')} \right)_{R_j^e, R_{j'}^e} \qquad [4.14]$$

The $3N$ equations of motion of the crystal lattice are written as:

$$M_j \ddot{u}_\alpha(j) + \sum_{\beta,j'} A_{\alpha\beta}(j,j')u_\beta(j') = 0 \qquad [4.15]$$

In the approximation of the binary potential energy, the application of the symmetry operations of the group O_h [MAR 65, MAN 68, MAN 71], such as the space inversion, allows certain properties of the force constants to be obtained, such as:

$$\begin{aligned} A_{\alpha\beta}(j',j') &= -\sum_{j\neq j'} A_{\alpha\beta}(j,j') \\ A_{\alpha\beta}(j,j') &= A_{\beta\alpha}(j',j) \end{aligned} \qquad [4.16]$$

The set of properties will be very useful to study the case of a molecule trapped in the matrix as well as its dynamics.

Furthermore, assuming that the dynamic variable harmonically depends on time, $u_\alpha(j)$ can be written in the form:

$$u_\alpha(j) = u_\alpha^0(j)e^{i\omega t} \qquad [4.17]$$

The dynamic matrix $D_{\alpha\beta}^{\mathrm{C}}(\vec{K})$ of the lattice is defined in the absolute frame by:

$$D_{\alpha\beta}^{\mathrm{C}}(\vec{K}) = \frac{1}{M}\sum_j A_{\alpha\beta}(0,j)e^{-i\vec{K}.\vec{R}_j^{\mathrm{e}}} \qquad [4.18]$$

where $\vec{R}_j^{\mathrm{e}}$ and $\vec{K}$ represent, respectively, the equilibrium position vector of the atom j with respect to the origin of the frame and the wave vector. M is the mass of an atom of the lattice.

The eigenvalues and eigenvectors of this dynamic matrix are obtained by solving the equations:

$$\sum_\beta D_{\alpha\beta}^{\mathrm{C}}(\vec{K})\sigma_\beta^l(\vec{K}) = \omega_l^2(\vec{K})\sigma_\alpha^l(\vec{K}) \qquad [4.19]$$

For a wave vector $\vec{K}$, the eigenvalues are the characteristic frequencies $\omega_l^2(\vec{K})$ of the bands of phonons and the eigenvectors are the polarization vectors $\vec{\sigma}^l(\vec{K})$, orthonormal complex vectors verifying the conditions:

$$\begin{aligned} &\sigma_\alpha^{l\,*}(\vec{K})\sigma_\alpha^{l'}(\vec{K}) = \delta_{l,l'} \\ &\sigma_\alpha^{l\,*}(\vec{K})\sigma_\beta^{l}(\vec{K}) = \delta_{\alpha,\beta} \end{aligned} \qquad [4.20]$$

where l designates the three acoustic frequency bands determined by the dispersion relations of the crystal.

We introduce the vibrational normal coordinates $q^l(\vec{K})$ (complex variables) by the relation:

$$u_\alpha(j) = \frac{1}{\sqrt{NM}} \sum_{l,K} q^l(\vec{K})\sigma_\alpha^l(\vec{K})e^{i\vec{K}.\vec{R}_j^e} \qquad [4.21]$$

which allows the crystal's Hamiltonian to be written in the canonical form:

$$H_C = \frac{1}{2}\sum_{l,K}\left[\dot{q}^l(\vec{K})\dot{q}^l(-\vec{K}) + \omega_l^2(\vec{K})q^l(\vec{K})q^l(-\vec{K})\right] \qquad [4.22]$$

from which the energy levels are obtained as:

$$E_C = \sum_{l,K}\left(n_l(\vec{K}) + \frac{1}{2}\right)\hbar\omega_l(\vec{K}) \qquad [4.23]$$

The set of vibrational frequencies (phonons) of the lattice is obtained by solving equation [4.15]. The number N of atoms in a crystal is extremely large, and consequently the $3N$ eigenvalues form a continuum spectrum. Thus, the density of phonon states $\nu(\omega)$ is defined as the number of modes whose frequency is located in the interval $[\omega, \omega + d\omega]$. As an example, Figure 4.4 shows the normalized density of phonon states curve $\omega^2\nu(\omega^2)$ as well as the function $S(\omega^2)$ of the perfect argon crystal [MAN 71], which will be defined later.

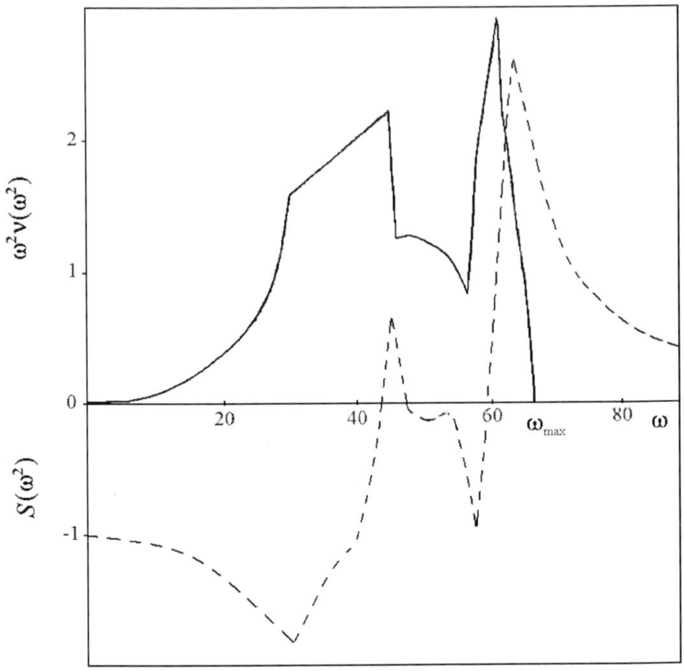

Figure 4.4. *Density of phonon states $\omega^2\nu(\omega^2)$ and function $S(\omega^2)$ curves of the perfect argon crystal*

4.3.3. *Green function of the perfect monoatomic crystal*

When a point defect (impurity) is introduced in a single substitutional site, it replaces an atom of the crystal. A mass defect and a force constant defect then result. The density of the phonon states is then found to be modified.

In order to solve the equations of motion of the doped lattice, it is practical to introduce the Green function matrix $\boldsymbol{G}^C$ of the perfect crystal, with elements $G^C_{\alpha\beta}$ corresponding to the directions α and β in the fixed frame:

$$G^C_{\alpha\beta}(j,j';\omega^2)=\frac{1}{NM}\sum_{l,K}\frac{\sigma^{l*}_{\alpha}(\vec{K})\sigma^{l}_{\beta}(\vec{K})}{\omega_l^2(\vec{K})-\omega^2}e^{i\vec{K}.\left(\vec{R}^e_{j'}-\vec{R}^e_j\right)} \qquad [4.24]$$

These elements of the Green function matrix are invariant by translation; they only depend on the relative distance between the atoms:

$$G^{C}_{\alpha\beta}(j,j';\omega^2) = G^{C}_{\alpha\beta}(j-j',0;\omega^2) = G^{C}_{\alpha\beta}(j-j';\omega^2) \quad [4.25]$$

and fulfill the properties of the octahedral symmetry group O_h.

Table 4.4 shows the numerical values of some elements of the (dimensionless) Green function matrix between an atom j and its neighbors, according to their relative distance [FLI 62].

We note that the values of these elements following two different directions are negligible. Moreover, the values of the other elements rapidly diminish when the relative distance between the atoms increases.

Distance	g_{XX}	g_{YY}	g_{ZZ}	g_{XY}	g_{XZ}	g_{YZ}
0	0.838	0.838	0.838	0	0	0
$a/\sqrt{2}$	0.257	0.257	0.206	0	0	0.081
a	0.145	0.143	0.143	0	0	0
$a\sqrt{2}$	0.144	0.128	0.128	0.016	0.027	0.027

Table 4.4. *Numerical values of some elements of the (dimensionless) Green function matrix of a perfect monoatomic crystal, as a function of the relative distance between an atom j and an atom j' nearest, second and third neighbor*

4.4. Inclusion of a molecule in a rare gas matrix

4.4.1. *Deformation method*

In a nanocage assumed to be at temperature zero, the inclusion of a diatomic molecule causes a more or less significant distortion depending on the nature and size of the molecule and the nanocage.

In a rare gas matrix, the included molecule can substitute one or two (or more) of the matrix's neighboring atoms. However, for molecules with a reasonable size (such as HCl, CO and N_2), we will consider only the substitution sites: "single S_1" or "double S_2" as shown in Figure 4.5.

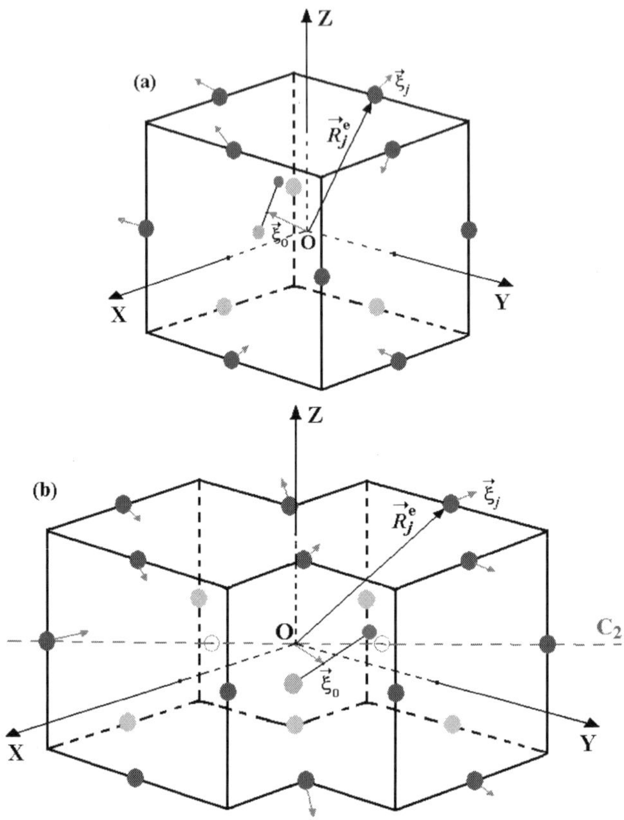

Figure 4.5. *Inclusion sites of a diatomic molecule in a rare gas matrix: a) single substitutional site and b) double substitutional site. $\vec{R}_j^e$ is the equilibrium position vector of the atom j in the fixed frame $(O,\vec{X},\vec{Y},\vec{Z})$ tied to the crystal and C_2 is a second-order symmetry axis. For a color version of this figure, see www.iste.co.uk/dahoo/infrared.zip*

The interaction potential energy of the doped crystal "trapped molecule-matrix" is given by:

$$U\left(\left\{\vec{R}_j^{\mathrm{e}},\vec{\xi}_j\right\},\vec{\xi}_0,\vec{\Omega}_0,q\right)=V_{\mathrm{C}}+\hat{V}_{\mathrm{MC}}\left(\left\{\vec{R}_j^{\mathrm{e}},\vec{\xi}_j\right\},\vec{\xi}_0,\vec{\Omega}_0,q\right)+V_{\mathrm{vib}}(q) \qquad [4.26]$$

where $\left(\left\{\vec{R}_j^{\mathrm{e}},\vec{\xi}_j\right\}\right)$ is the set of position vectors of the matrix atoms in the perfect crystal and their displacement vector due to the molecule's inclusion

$\vec{\xi}_0$, $\vec{\Omega}_0 \equiv (\phi_0, \theta_0)$ and q, respectively, represent the displacement vector, orientation and internal vibration of the molecule in its inclusion site. On other hand, V_{C} (equation [4.13]) and $V_{\mathrm{vib}}(q)$ (Chapter 2, equation [2.43]) are the potential energies of the perfect crystal and the internal vibration of molecule in the gaseous phase, and $\hat{V}_{\mathrm{MC}}$ represents the interaction energy molecule-matrix from which the interaction energy between the substituted atom or atoms and the rest of the matrix outside the molecule has been subtracted:

$$\begin{aligned}\hat{V}_{\mathrm{MC}}\left(\left\{\vec{R}^{\mathrm{e}}_{0j},\vec{\xi}_j\right\},\vec{\xi}_0,\vec{\Omega}_0,q\right) &= \sum_i\sum_j 4\varepsilon_{ij}\left\{\left(\frac{\sigma_{ij}}{\left|\vec{\mathrm{r}}_{ij}\right|}\right)^{12}-\left(\frac{\sigma_{ij}}{\left|\vec{\mathrm{r}}_{ij}\right|}\right)^{6}\right\} \\ &-\frac{1}{2}\vec{E}_{0\to j}.\vec{\vec{\alpha}}_j.\vec{E}_{0\to j}-\sum_j 4\varepsilon_{\mathrm{gr-gr}}\left\{\left(\frac{\sigma_{\mathrm{gr-gr}}}{\left|\vec{R}^{\mathrm{e}}_{0j}\right|}\right)^{12}-\left(\frac{\sigma_{\mathrm{gr-gr}}}{\left|\vec{R}^{\mathrm{e}}_{0j}\right|}\right)^{6}\right\}\end{aligned} \quad [4.27]$$

In this equation: $\vec{\vec{\alpha}}_j$ and $\vec{E}_{0\to j}$ are, respectively, the dipolar polarizability tensor of the atom j of the rare gas and the field created by the multipole moments of the molecule at this atom's place; $\vec{\mathrm{r}}_{ij}$ is the distance vector between the atom i of the molecule and the atom j of the matrix; and ε and σ are the LJ potential parameters.

The dependence on the molecule's angular and vibrational coordinates is contained in the position vector $\vec{\mathrm{r}}_i$ of the atom i of the molecule through the distance vector $\vec{\mathrm{r}}_{ij}$ as:

$$\vec{\mathrm{r}}_{ij} = \vec{\mathrm{r}}_{0j} - \vec{\xi}_0 - \vec{\mathrm{r}}_i = \vec{R}^{\mathrm{e}}_{0j} + \vec{\xi}_j - \vec{\xi}_0 - \vec{\mathrm{r}}_i \quad [4.28]$$

as well as in its multipole moments. A Taylor series expansion of these characteristics as a function of the vibrational coordinate allows one to write:

$$\vec{\mathrm{r}}_i = \vec{\mathrm{r}}_i^{\mathrm{e}} + \left(\frac{\partial \vec{\mathrm{r}}_i}{\partial q}\right)_{q=0} q + \frac{1}{2}\left(\frac{\partial^2 \vec{\mathrm{r}}_i}{\partial q^2}\right)_{q=0} q^2 + \ldots \quad [4.29a]$$

$$\vec{\mu}_0 = \vec{\mu}_0^{\mathrm{e}} + \left(\frac{\partial \vec{\mu}_0}{\partial q}\right)_{q=0} q + \frac{1}{2}\left(\frac{\partial^2 \vec{\mu}_0}{\partial q^2}\right)_{q=0} q^2 + \ldots \qquad [4.29b]$$

$$\vec{\vec{\Theta}}_0 = \vec{\vec{\Theta}}_0^{\mathrm{e}} + \left(\frac{\partial \vec{\vec{\Theta}}_0}{\partial q}\right)_{q=0} q + \frac{1}{2}\left(\frac{\partial^2 \vec{\vec{\Theta}}_0}{\partial q^2}\right)_{q=0} q^2 + \ldots \qquad [4.29c]$$

These molecular characteristics are defined in the frame $(G,\vec{x},\vec{y},\vec{z})$ tied to the molecule. Their explicit dependence on the angular coordinates (ϕ_0,θ_0) appears when they are expressed in the fixed frame $(O,\vec{X},\vec{Y},\vec{Z})$ by using the transformation matrix $\mathbf{M}(\phi_0,\theta_0)$ from the fixed frame to the frame tied to the molecule (section 4.8.2) [ROS 67].

4.4.2. *Equilibrium of the doped crystal*

The equilibrium of the doped crystal is obtained when its total internal force is zero. In other words, the total energy U is minimal:

$$\vec{\nabla} U\left(\left\{\vec{R}_{0j}^{\mathrm{e}},\vec{\xi}_j\right\},\vec{\xi}_0,\vec{\Omega}_0,q\right)=\vec{0} \qquad [4.30]$$

The gradient operator $\vec{\nabla}$ affects all of the system's degrees of freedom. In fact, the internal vibration movement of the included molecule is much faster than its external movements and those of the matrix's atoms. As a result, we will be able to determine the distortion of the matrix around the molecule assumed to be rigid.

In the approximation of a reasonably small distortion of the matrix atoms around the trapped molecule, the total interaction energy U can be expanded as a Taylor series up to the quadratic order in terms of the displacement vectors $\vec{\xi}$ of the atoms and the molecule's center of mass. It is written in the following general form:

$$U = V_{\mathrm{vib}}(q) + V_{\mathrm{C}} + \frac{1}{2}\vec{\xi}.\vec{\nabla}\vec{\nabla}V_{\mathrm{C}}.\vec{\xi} + \hat{V}_{\mathrm{MC}} + \vec{\nabla}\hat{V}_{\mathrm{MC}}.\vec{\xi} + \frac{1}{2}\vec{\xi}.\vec{\nabla}\vec{\nabla}\hat{V}_{\mathrm{MC}}.\vec{\xi} \qquad [4.31]$$

The application of the equilibrium condition of the doped crystal and the introduction of the Green function matrix of the perfect crystal defined in equation [4.24] [FLI 62], make it possible to determine the displacement vectors $\vec{\xi}_k$ due to the inclusion of the molecule ($k = 0$ for the molecule's center of mass and $k = j$ for the matrix's atoms). They are given by the iterative expression:

$$\vec{\xi}_k = \sum_{k'} \mathrm{G^C}(k,k')\vec{F}(k') + \sum_{k',k''} \mathrm{G^C}(k',k'')\mathrm{H}(k'',k')\vec{\xi}_{k'} + \ldots, \qquad [4.32]$$

where $\vec{F}$ and H are, respectively, the force vector and the force constant matrix applied to the atoms and the molecule after its inclusion and $\mathrm{G^C}$ are the Green function matrix of the perfect crystal defined previously. For the face-centered cubic lattice, the values of this matrix's elements were determined for the first layers around the inclusion site [FLI 62] and they show a quasi-isotropic symmetry. However, this does not imply an isotropic distortion, because the tensors $\vec{F}$ and H are fundamentally anisotropic and depend both on the size and shape of the molecule and on the kind of site, both single or double substitutional site.

The displacement vectors in the matrix are determined by substituting the solution of the zeroth-order $\vec{\xi}_k$ in the second term of equation [4.32]. Approximatively, we obtain:

$$\vec{\xi}_k \approx \sum_{k'} \mathrm{G^C}(k,k')\left(\vec{F}(k') + \sum_{k'',k'''} \mathrm{H}(k',k'')\mathrm{G^C}(k'',k''')\vec{F}(k''')\right) \qquad [4.33]$$

These displacement vectors are functions of the angular (ϕ_0, θ_0) and vibrational q coordinates of the molecule in its site through the forces' and force constants' tensors. However, by using the adiabatic approximation, it is possible to determine $\vec{\xi}_k$ for each set of coordinates (q, ϕ_0, θ_0) and, thus, write it as:

$$\vec{\xi}_k \approx \vec{\xi}_k^{\,s} + \vec{\xi}_k(q, \vec{\Omega}_0) \qquad [4.34]$$

where $\vec{\xi}_k^{\,s}$ represents the displacement vector of the molecule's center of mass ($k = 0$) and the atoms of the doped crystal ($k = j$) obtained by minimizing the total energy (static model or cage model), given that the molecule is rigid and at its orientational equilibrium, and $\vec{\xi}_k(q,\vec{\Omega}_0)$ represents the variation due to the vibrational and orientational dependence about $\vec{\xi}_k^{\mathrm{s}}$.

It is evident that the equilibrium condition (equation [4.28]) is only verified when the complete expression of $\vec{\xi}_k$ is taken into account. However, if the matrix practically does not distort with the molecule's vibration and orientation dynamics, then the displacement vectors $\vec{\xi}_k$ may be approximated by $\vec{\xi}_k^{\mathrm{s}}$. In the opposite case, which would correspond to a more or less significant modulation of $\vec{\xi}_k$ depending on the variables $(q,\vec{\Omega}_0)$, the second part is no longer negligible and it is necessary to process the Hamiltonian differently.

4.5. General Hamiltonian and separation of the movements

4.5.1. *Hamiltonian of the system*

The Hamiltonian of the system constituted by the trapped molecule and the matrix is written as:

$$H = T_{\mathrm{vib}} + T_{\mathrm{rot}} + T_{\mathrm{trans}} + T_{\mathrm{C}} + U \qquad [4.35]$$

with

$$T_{\mathrm{trans}} = \sum_{\alpha} \frac{P_{\alpha}^2(0)}{2M_0} \qquad [4.36]$$

T_{vib}, T_{rot} and T_{trans}, respectively, designate the kinetic Hamiltonians of the molecule's vibration (equation [2.18]), rotation (equation [2.82]) and translation, and T_{C} is the lattice's vibration kinetic Hamiltonian (equation [4.13]). Let $\vec{\zeta}$ be the lattice's vibration dynamic variable defined from the equation of the perfect crystal. We then express $\vec{\zeta}$ in the form of a sum of the variable $\vec{\xi}$ defined by equation [4.33] and the small amplitude dynamic variable $\vec{u}$, such that:

$$\vec{\zeta} = \vec{\xi} + \vec{u} \qquad [4.37]$$

The interaction potential energy U, defined by equation [4.24], can moreover be put in the generalized symbolic form:

$$U = \phi^{(0)}(q,\vec{\Omega}_0) + \underline{\phi}^{(1)}(q,\vec{\Omega}_0).\vec{u} + \frac{1}{2}\underline{\phi}^{(2)}(q,\vec{\Omega}_0)\vec{u}.\vec{u} + ... \qquad [4.38]$$

Such a formulation means recentering the movements around an equilibrium configuration of a "prepared" crystal comprising the molecule and not around that of the perfect crystal.

The first term $\varphi^{(0)}(q,\vec{\Omega}_0)$ of the right-hand side of equation [4.38] is the independent potential energy of the lattice's dynamics, and it accounts for a significant part of the dynamics of the molecule's vibration–orientation. It is written as:

$$\begin{aligned}\varphi^{(0)}(q,\vec{\Omega}_0) = V_{\text{vib}}(q) + \hat{V}_{\text{MC}}(q,\vec{\Omega}_0) + \vec{\nabla}\hat{V}_{\text{MC}}(q,\vec{\Omega}_0).\vec{\xi} \\ + \frac{1}{2}\vec{\nabla}\vec{\nabla}\left(V_{\text{C}} + \hat{V}_{\text{MC}}(q,\vec{\Omega}_0)\right).\vec{\xi}.\vec{\xi} + ...\end{aligned} \qquad [4.39]$$

The second term $\underline{\varphi}^{(1)}(q,\vec{\Omega}_0)$, which is linear in $\vec{u}$, characterizes the forces acting on the molecule and on the atoms of the crystal and is given by:

$$\underline{\varphi}^{(1)}(q,\vec{\Omega}_0) = \vec{\nabla}\hat{V}_{\text{MC}}(q,\vec{\Omega}_0) + \vec{\nabla}\vec{\nabla}\left(V_{\text{C}} + \hat{V}_{\text{MC}}(q,\vec{\Omega}_0)\right).\vec{\xi} + ... \qquad [4.40]$$

This term is zero, at this order of expansion in $\vec{\xi}$, if $\vec{\xi}$ verifies equation [4.33], but it gives a non-zero contribution if only the part $\vec{\xi}_k^{\text{s}}$, independent of the molecule's dynamics, is taken into account.

The third term $\underline{\varphi}^{(2)}(q,\vec{\Omega}_0)$ represents the harmonic force constants:

$$\underline{\varphi}^{(2)}(q,\vec{\Omega}_0) = \vec{\nabla}\vec{\nabla}\left[V_{\text{C}} + \hat{V}_{\text{MC}}(q,\vec{\Omega}_0) + \vec{\nabla}\left(V_{\text{C}} + \hat{V}_{\text{MC}}(q,\vec{\Omega}_0)\right).\vec{\xi} + ...\right] \qquad [4.41]$$

As for the terms of the higher orders, they represent the anharmonic force constants.

It should be noted that the influence of the matrix's distortion on the force constants is only accounted for by the linear term in $\vec{\xi}$. Canceling out this term in $\underline{\varphi}^{(1)}(q,\vec{\Omega}_0)$ and in the terms of the higher orders is equivalent to considering that the force constants of the matrix's atoms are those of a non-distorted crystal despite the presence of the included molecule. Moreover, $\underline{\varphi}^{(1)}(q,\vec{\Omega}_0)$ depend on the molecular variables by both the distortion $\vec{\xi}_k(q,\vec{\Omega}_0)$ and the interaction energy $\hat{V}_{\text{MC}}(q,\vec{\Omega}_0)$.

The total Hamiltonian can then be decomposed as:

$$H=\left[T_{\text{vib}}+T_{\text{rot}}+\varphi^{(0)}(q,\vec{\Omega}_0)\right]+\left[T_{\text{trans}}+T_{\text{C}}+\sum_{l\neq 0}\underline{\varphi}^{(l)}(q,\vec{\Omega}_0).\vec{u}^{l}\right] \quad [4.42]$$

The first part of this expression is the Hamiltonian relative to the movements of what we will from now on call "the optical system or the optically active molecule" characterized by the vibration and orientation variables of the trapped molecule. The second part of the expression [4.42] designates the Hamiltonian of the "bath" characterized by the movements of the molecule's center of mass, the lattice vibrations "phonons" and the dynamic part of the potential energy coupling the variables $(q,\vec{\Omega}_0)$ and $\vec{u}$. We note that the movement of the molecule's center of mass can have a vibrational frequency in the band of phonons of the perfect crystal, which correspond to a non-localized phonon. If the frequency is greater than the cut-off frequency of this band, we then refer to the localized mode.

A judicious separation of this Hamiltonian now requires the dynamic conditions to be specified in order to solve the equations of motion.

4.5.2. *Separation of the optical system's movements and the bath in the rigid matrix approximation*

Assuming that the matrix is not dragged by the molecule's vibration and orientation movements, either because the matrix's rigidity is sufficiently large with respect to the forces binding the molecule to the matrix or because the movements of this molecule are fast enough to annihilate, on average, the dragging effect is accounted for by the approximation $\vec{\xi}(q,\vec{\Omega}_0) \cong \vec{\xi}^{\mathrm{s}}$ and it is then possible to obtain a separation of the total Hamiltonian in the form:

$$H = \tilde{H}_{\mathrm{vib}} + \tilde{H}_{\mathrm{orient}} + \tilde{H}_{\mathrm{bath}} + \Delta\tilde{H}(q,\vec{\Omega}_0,\vec{u}) \qquad [4.43]$$

4.5.3. *Vibrational mode*

In equation [4.43], the first term $\tilde{H}_{\mathrm{vib}}$ is the renormalized vibration Hamiltonian of the molecule in its inclusion site:

$$\tilde{H}_{\mathrm{vib}} = H^0_{\mathrm{vib}} + \Delta V^{\mathrm{anhar}}_{\mathrm{vib}}(q) + V^{\mathrm{e}}_{\mathrm{MC}}(q) \qquad [4.44]$$

where H^0_{vib} and $\Delta V^{\mathrm{anhar}}_{\mathrm{vib}}(q)$ are, respectively, the vibration Hamiltonian (harmonic oscillator) and the anharmonic correction of the potential energy function of the molecule in the gaseous phase (Chapter 2, equations [2.43] and [2.44]), and $V^{\mathrm{e}}_{\mathrm{MC}}(q)$ represents the perturbation at the potential energy of the molecule in its inclusion site taken at its equilibrium orientation $(\phi^{\mathrm{e}}_0, \theta^{\mathrm{e}}_0)$. The experimental studies show that the vibrational frequency shifts are generally small compared to these frequencies. The series expansion of $V^{\mathrm{e}}_{\mathrm{MC}}(q)$ can be limited to the second order:

$$V^{\mathrm{e}}_{\mathrm{MC}}(q) = \left(\frac{\partial V^{\mathrm{e}}_{\mathrm{MC}}}{\partial q}\right)_{q=0} q + \frac{1}{2}\left(\frac{\partial^2 V^{\mathrm{e}}_{\mathrm{MC}}}{\partial q^2}\right)_{q=0} q^2 = \beta_1 q + \frac{1}{2}\beta_2 q^2 \qquad [4.45]$$

The quadratic term $\frac{1}{2}\beta_2 q^2$ of this expression represents a perturbation of the Hamiltonian H^0_{vib} of the molecule's harmonic vibration in the gaseous phase and thus induces a variation of the harmonic force constant k.

The angular frequency of the vibration movement $\omega_0 = \sqrt{\frac{k}{\mu}}$ (equation [2.11]) becomes $\omega_0^{'} = \sqrt{\frac{k+\beta_2}{\mu}}$. The eigenenergies of the Hamiltonian of the new harmonic oscillator are then written as:

$$E_{\mathrm{v}}^{0'} = \left(\mathrm{v}+\frac{1}{2}\right)\hbar\omega_0^{'} = \left(\mathrm{v}+\frac{1}{2}\right)h\nu_0^{'} \qquad [4.46]$$

On the other hand, introducing the linear term $\beta_1 q$ of equation [4.45], through a variable transformation, enables the Hamiltonian to be written in the form:

$$H_{\mathrm{vib}}^{'} = -\frac{1}{2}\hbar\omega_0^{'}\left(\frac{\partial^2}{\partial q'^2} - q'^2\right) - \frac{\beta_1^2}{2\hbar\omega_0^{'}} \qquad [4.47]$$

in which q' is the new normal coordinate of the shifted harmonic oscillator. It is given by:

$$q' = q + \frac{\beta_1}{\hbar\omega_0^{'}} \qquad [4.48]$$

The eigenenergies of the Hamiltonain $H_{\mathrm{vib}}^{'}$ are therefore all shifted by the same quantity, such as:

$$E_{\mathrm{v}}^{'} = \left(\mathrm{v}+\frac{1}{2}\right)\hbar\omega_0^{'} - \frac{\beta_1^2}{2\hbar\omega_0^{'}} \qquad [4.49]$$

Finally, the anharmonic part $\Delta V_{\mathrm{vib}}^{\mathrm{anhar}}(q)$ of the potential energy function can be written according to q' as:

$$\Delta V_{\mathrm{vib}}^{\mathrm{anhar}}(q') = k_3^{'}q'^3 + k_4^{'}q'^4 + k_5^{'}q'^5 \qquad [4.50]$$

and is introduced in the same way as in Chapter 2 where we replace the constants k_3, k_4 and k_5 by $k_3^{'}$, $k_4^{'}$ and $k_5^{'}$. We then obtain new expressions for the eigenvalues and the eigenvectors (equations [2.47]–[2.51]). The vibrational frequency shifts $\Delta\omega_{\mathrm{vv'}}$ between two levels v and v' of the molecule trapped in the nanocage can then be determined.

4.5.4. *Orientational modes*

In equation [4.41], $\tilde{H}_{\mathrm{orient}}$ is the renormalized Hamiltonian of the orientation movement of the rigid molecule trapped in the nanocage at the equilibrium position of its center of mass. It is written as:

$$\tilde{H}_{\mathrm{orient}} = T_{\mathrm{rot}} + V_{\mathrm{MC}}^{\mathrm{e}}(\phi_0, \theta_0) \qquad [4.51]$$

in which T_{rot} is the kinetic rotation energy operator of the rigid, free molecule (Chapter 2, equation [2.82]) and $V_{\mathrm{MC}}^{\mathrm{e}}(\phi_0, \theta_0)$ is the perturbed orientational potential energy surface of the molecule in its inclusion site.

The solution of the Schrödinger equation associated with the Hamiltonian $\tilde{H}_{\mathrm{orient}}$:

$$\tilde{H}_{\mathrm{orient}} \left| jm \right\rangle = \left[T_{\mathrm{rot}} + V_{\mathrm{MC}}^{\mathrm{e}}(\phi_0, \theta_0) \right] \left| jm \right\rangle = E_{jm} \left| jm \right\rangle \qquad [4.52]$$

allows the eigenenergies E_{jm} and eigenvectors $\left| jm \right\rangle$ of the perturbed orientation movement to be determined, in which j and m are the quantum numbers, respectively, associated with the angular variables of nutation θ_0 and precession ϕ_0.

The techniques used to solve equation [4.52] fundamentally depend on the shape of the potential energy surface $V_{\mathrm{MC}}^{\mathrm{e}}(\phi_0, \theta_0)$ of the rigid molecule around its equilibrium orientation in its inclusion site, particularly the perturbation amplitude of the rotation movement.

4.5.4.1. *Weakly perturbed rotation*

When the diatomic molecule can perform a more or less perturbed rotation movement, the potential energy $V_{\mathrm{MC}}^{\mathrm{e}}(\phi_0, \theta_0)$ can be analytically developed or numerically adjusted in the basis of the spherical harmonics $Y_{l_0}^{p_0}(\theta_0, \phi_0)$ in the form:

$$V^{\mathrm{e}}_{\mathrm{MC}}(\phi_0,\theta_0)=\sum_{l_0=1}^{l_{\max}}\sum_{p_0=-l_0}^{+l_0} A^{l_0}_{p_0}\times Y^{p_0}_{l_0}(\theta_0,\phi_0) \qquad [4.53]$$

In this expression, $A^{l_0}_{p_0}$ are complex coefficients of the expansion, l_0 is an integer number associated with the angular coordinate of nutation θ_0 and p_0 $(=-l_0,\dots,0,\dots,+l_0)$ is associated with the angular coordinate of precession ϕ_0. The values taken by the number l are such that finally the expression [4.53] is the best representation of the potential energy surface $V^{\mathrm{e}}_{\mathrm{MC}}(\phi_0,\theta_0)$.

To solve the eigenvalues equation [4.52], we use the large perturbation technique by writing the eigenvector $|jm\rangle$ as a linear combination of the eigenvectors $|JM\rangle$ of the molecule's free rotation in the gaseous phase, such that:

$$|jm\rangle=\sum_{J,M}\Lambda^{jm}_{JM}|JM\rangle \qquad [4.54]$$

in which Λ^{jm}_{JM} are complex coefficients determined by solving equation [4.52] and $|JM\rangle$ are solutions of the equation:

$$T_{\mathrm{rot}}|JM\rangle=E_J|JM\rangle=B^e J(J+1)|JM\rangle \qquad [4.55]$$

By introducing equations [4.53]–[4.55] into the eigenvalues equation [4.52], there comes:

$$\sum_{J,M}\Lambda^{jm}_{JM}\left[\left(E_J-E_{jm}\right)+\sum_{l_0,p_0}A^{l_0}_{p_0}\times Y^{p_0}_{l_0}(\theta_0,\phi_0)\right]|JM\rangle=0 \qquad [4.56]$$

The secular equation is obtained by multiplying equation [4.56] to the left-hand side by the bra $\langle J'M'|$ and using equation [2.89] (Chapter 2), so that:

$$\sum_{J,M}\left[\left(E_{J'}-E_{jm}\right)\delta_{JJ'}\delta_{MM'}+\sum_{l_0,p_0}A_{p_0}^{l_0}\times\left\langle J'M'\right|Y_{l_0}^{p_0}(\theta_0,\phi_0)\left|JM\right\rangle\right]\Lambda_{JM}^{jm}=0 \quad [4.57]$$

where δ is the Kronecker symbol and where the matrix element $\left\langle J'M'\right|Y_{l_0}^{p_0}(\theta_0,\phi_0)\left|JM\right\rangle$ can be expressed in terms of the Clebsch–Gordan coefficients as:

$$\left\langle J'M'\right|Y_{l_0}^{p_0}(\theta_0,\phi_0)\left|JM\right\rangle=\sqrt{\frac{(2J+1)}{(2J'+1)}}C(Jl_0J';Mp_0M')C(Jl_0J';000) \quad [4.58]$$

Thus, the secular equation becomes:

$$\sum_{J,M}\left[\begin{array}{l}\left(E_{J'}-E_{jm}\right)\delta_{JJ'}\delta_{MM'}+\sum_{l_0,p_0}A_{p_0}^{l_0}\times\sqrt{\dfrac{(2J+1)}{(2J'+1)}}\\ \qquad\times C(Jl_0J';Mp_0M')C(Jl_0J';000)\end{array}\right]\Lambda_{JM}^{jm}=0 \quad [4.59]$$

and the determination of the eigenvectors and the eigenvalues of the molecule's rotation is reduced to diagonalizing a square matrix whose matrix elements $M_{J'M'}^{JM}$ are given by:

$$\begin{array}{l}M_{J'M'}^{JM}=\left(E_{J'}-E_{jm}\right)\delta_{JJ'}\delta_{MM'}+\sqrt{\dfrac{(2J+1)}{(2J'+1)}}\times\\ \qquad\sum_{l_0,p_0}A_{p_0}^{l_0}\times C(Jl_0J';Mp_0M')C(Jl_0J';000)\end{array} \quad [4.60]$$

COMMENT.– The Clebsch–Gordan coefficients are only non-zero if $\left|J-l_0\right|\leq J'\leq J+l_0$ and $M'=M+p_0$.

The dimension of this matrix depends on the maximum value taken by the number l_0 in the expression [4.53] and the rotational energy levels implied in the spectral lines. Indeed, the levels j and j' are perturbed (they are contaminated) by each other through the number l_0. For example, if $l_{max}=5$, then a level j may be perturbed by a level labeled $j+5$.

4.5.4.1.1. Libration

Analyzing the potential energy surface $V_{\mathrm{MC}}^{\mathrm{e}}(\phi_0,\theta_0)$ determined numerically shows that the molecule can only perform angular oscillation movements (librations) with a more or less large amplitude around an equilibrium configuration $(\phi_0^{\mathrm{e}},\theta_0^{\mathrm{e}})$. Several cases may arise, as discussed in the following.

4.5.4.1.2. First case

The molecule performs a one-dimensional libration movement around the equilibrium configuration $\theta_0^{\mathrm{e}}=0$, the precession movement in ϕ_0 (around the OZ axis of the fixed frame) being quasi-free. The angular coordinate θ_0 is no longer suitable for describing the libration movement. It is therefore necessary to introduce a new angular coordinate $t_\theta=\tan(\theta_0/2)$, which enables small variations near 0 to be described.

Consequently, the potential energy $V_{\mathrm{MC}}^{\mathrm{e}}(\phi_0,\theta_0)$ can be adjusted to a series of successive powers of t_θ and depending on ϕ_0 such that:

$$V_{\mathrm{MC}}^{\mathrm{e}}(\phi_0,t_\theta)=\sum_{\gamma=1}K_\gamma\times t_\theta^\gamma\times F_\gamma(\phi_0) \qquad [4.61]$$

where $F_\gamma(\phi_0)$ are functions close to unity, given the weak dependence of the energy in ϕ_0 (quasi-free movement) and where K_γ are constants determined by the least squares method, with $K_0=V_{\mathrm{MC}}^{\mathrm{e}}(\phi_0^{\mathrm{e}},\theta_0^{\mathrm{e}})$ being the energy value for the equilibrium configuration.

However, given the quasi-free rotation in ϕ_0, the libration movement is quasi-symmetric; the constants related to the odd powers of T_{rot} are small and may be ignored. Moreover, for oscillations with weak amplitude, the constant K_2 of the harmonic term (second order) is much larger than those of higher orders; and in the approximation of the free rotation in ϕ_0, the functions $F_\gamma(\phi_0)$ can be assumed to be constant and equal to 1. The expression [4.61] then becomes:

$$V_{\mathrm{MC}}^{\mathrm{e}}(\phi_0, t_\theta) = K_2 t_\theta^2 + K_4 t_\theta^4 + K_6 t_\theta^6 + \dots \qquad [4.62]$$

In order to transform the kinetic rotation energy operator T_{rot} (equation [2.82] of Chapter 2) into a libration operator, we transform the variable in the following way:

$$u_\theta = 2\sqrt{\alpha} t_\theta$$

in which $\alpha = \frac{1}{2}\sqrt{\frac{K_2}{B^e}}$

K_2 is the harmonic constant of equation [4.62] and B^e is the rotational constant of the molecule. The kinetic rotation energy operator is now written as:

$$\begin{aligned} T_{\mathrm{rot}} &\approx -\alpha B^e \left\{ T_{u_\theta} + \frac{1}{2\alpha} u_\theta^2 T_{u_\theta} + \frac{1}{16\alpha^2} u_\theta^4 T_{u_\theta} \right\} \quad \text{with} \\ T_{u_\theta} &= \frac{\partial^2}{\partial u_\theta^2} + \frac{1}{u_\theta} \frac{\partial}{\partial u_\theta} + \frac{1}{u_\theta^2} \frac{\partial^2}{\partial \phi^2} \end{aligned} \qquad [4.63]$$

By substituting equations [4.62] and [4.63] into [4.51], the Hamiltonian $\tilde{H}_{\mathrm{orient}}$ can be rewritten as:

$$\begin{aligned} \tilde{H}_{\mathrm{orient}} &\approx -\alpha B^e \left(T_{u_\theta} - u_\theta^2 \right) \\ &+ \left\{ \begin{aligned} &-\frac{1}{2} B^e u_\theta^2 \left(T_{u_\theta} - u_\theta^2 \right) - \frac{1}{16\alpha} B^e u_\theta^4 \left(T_{u_\theta} - u_\theta^2 \right) \\ &+ \frac{1}{2}\left(\frac{K_4}{8\alpha^2} - B^e \right) u_\theta^4 + \frac{1}{16\alpha}\left(\frac{K_6}{4\alpha^2} - B^e \right) u_\theta^6 + \dots \end{aligned} \right\} \end{aligned} \qquad [4.64]$$

at the near constant value K_0. In this expression, the first term represents the harmonic order of the libration, and the following terms (in brackets) represent the anharmonic orders that are treated as perturbations.

Thus, in the harmonic oscillator basis, the eigenvalues and the eigenvectors associated with the harmonic term of $\tilde{H}_{\mathrm{orient}}$ are given by:

$$E_{jM}^{(0)} = \hbar\omega_\theta\left(2j - |M| + 1\right) \qquad \text{with} \quad \hbar\omega_\theta = 2\alpha B^e$$
$$|jM\rangle = N_j^{|M|}\exp\left(-u_\theta^2/2\right)u_\theta^{|M|}L_j^{|M|}\left(u_\theta^2\right)\exp\left(iM\phi_0\right) \qquad [4.65]$$
$$N_j^{|M|} = \left(\frac{\alpha}{\pi}\right)^{1/2}\left((j-|M|)!\right)^{1/2}(j!)^{-3/2}$$

where $N_j^{|M|}$ is a normalization factor, $L_j^{|M|}\left(u_\theta^2\right)$ designate the Laguerre associated functions and where the exponential term describes the free rotation in ϕ_0. The librational quantum number j is a natural integer and the rotation quantum number M is defined by $j \geq |M|$.

Finally, when the anharmonic terms of the expression [4.64] are introduced, by using the stationary perturbation theory to first-order energy, the eigenvalues can be cast as:

$$E_{jM} = E_{jM}^{(0)} + \Delta E_{jM} \qquad [4.66]$$

The expression of the corrective term ΔE_{jM} is given in section 4.8.3.

4.5.4.1.3. Second case

The molecule performs a two-dimensional libration movement around the equilibrium configuration (ϕ_0^e, θ_0^e) with $\theta_0^e \approx \pi/2$. The angular coordinates θ_0 and ϕ_0 are no longer suitable for describing the two libration movements. We introduce two new angular coordinates $u_\theta = \cos\theta_0$ and $u_\phi = \phi_0 - \phi_0^e$ (expressed in radians) varying around 0.

The potential energy $V_{\mathrm{MC}}^{\mathrm{e}}(\phi_0, \theta_0)$ can be adjusted to a series of successive powers in u_θ and u_ϕ in the form:

$$V_{\mathrm{MC}}^{\mathrm{e}}(u_\phi, u_\theta) = \sum_{\gamma,\lambda} K_{\gamma\lambda} \times u_\phi^\gamma \times u_\theta^\lambda \qquad [4.67]$$

in which $K_{00} = V_{\mathrm{MC}}^{\mathrm{e}}(\phi_0^{\mathrm{e}}, \theta_0^{\mathrm{e}})$ is the energy value for the equilibrium configuration.

However, because the oscillations of the two movements still have small amplitudes, the coefficients of the harmonic terms K_{20} and K_{02} are much larger than those of the other terms, such that the eigenvectors $|jm\rangle$ of the two-dimensional oscillator can be written as the tensor product of the eigenvectors $|j\rangle$ associated with the oscillator in θ_0 and $|m\rangle$ associated with that in ϕ_0 : $|jm\rangle = |j\rangle \otimes |m\rangle$.

By introducing the new angular coordinates into the expression of the kinetic rotation energy operator T_{rot}, we obtain:

$$T_{\text{rot}} \approx -B^e \left\{ \frac{\partial^2}{\partial u_\theta^2} - u_\theta^2 \frac{\partial^2}{\partial u_\theta^2} - 2u_\theta \frac{\partial}{\partial u_\theta} + \left(1 + u_\theta^2\right) \frac{\partial^2}{\partial u_\phi^2} \right\} \qquad [4.68]$$

By substituting equations [4.67] and [4.68] into [4.51], the Hamiltonian $\tilde{H}_{\text{orient}}$ can be rewritten, at the near constant value K_{00}, as:

$$\begin{aligned} \tilde{H}_{\text{orient}} \approx & -B^e \left\{ \frac{\partial^2}{\partial u_\theta^2} - \frac{K_{02}}{B^e} u_\theta^2 \right\} - B^e \left\{ \frac{\partial^2}{\partial u_\phi^2} - \frac{K_{20}}{B^e} u_\phi^2 \right\} \\ & + B^e \left\{ u_\theta^2 \frac{\partial^2}{\partial u_\theta^2} + 2u_\theta \frac{\partial}{\partial u_\theta} - u_\theta^2 \frac{\partial^2}{\partial u_\phi^2} + \frac{1}{B^e} \sum_{\gamma\lambda}^{*} K_{\gamma\lambda} \times u_\phi^\gamma \times u_\theta^\lambda \right\} \end{aligned} \qquad [4.69]$$

In this expression, the two first terms are the harmonic Hamiltonians associated with the two movements and the third term represents the perturbative Hamiltonian, which will be introduced using the stationary perturbations theory. In the latter term, the symbol * indicates that K_{20} and K_{02} are excluded.

Thus, in the harmonic approximation, the eigenvalues and the eigenvectors associated with the harmonic terms of $\tilde{H}_{\text{orient}}$ are given by:

$$\begin{aligned} & E_j^{(0)} = \hbar\omega_\theta \left(j + \tfrac{1}{2}\right) \qquad \text{with} \quad \hbar\omega_\theta = 2\sqrt{B^e K_{02}} \text{ and } \beta_\theta = \left(\frac{K_{02}}{B^e}\right)^{1/4} \\ & |j\rangle = N_j \exp\left(-\beta_\theta^2 u_\theta^2 / 2\right) H_j\left(\beta_\theta u_\theta\right) \\ & N_j = \left(\frac{\beta_\theta^2}{\pi}\right)^{1/4} \left(2^j j!\right)^{-1/2} \end{aligned} \qquad [4.70]$$

and

$$E_m^{(0)} = \hbar\omega_\phi\left(m+\tfrac{1}{2}\right) \qquad \text{with} \quad \hbar\omega_\phi = 2\sqrt{B^e K_{20}} \text{ and } \beta_\phi = \left(\frac{K_{20}}{B^e}\right)^{1/4}$$
$$|m\rangle = N_m \exp\left(-\beta_\phi^2 u_\phi^2/2\right) H_m\left(\beta_\phi u_\phi\right) \qquad [4.71]$$
$$N_m = \left(\frac{\beta_\phi^2}{\pi}\right)^{1/4}\left(2^m m!\right)^{-1/2}$$

where N are normalization factors and H designates the Hermite polynomials. The quantum numbers j and m are natural integers and are independent from each other.

If we make an adjustment to the potential energy $V_{\mathrm{MC}}^{\mathrm{e}}(u_\phi, u_\theta)$ up to the order $\gamma+\lambda=6$ and if we apply the stationary perturbation theory to first-order energy on the perturbative Hamiltonian of the expression [4.69], the expression of the eigenvalues can be written as:

$$E_{jm} = E_j^{(0)} + E_m^{(0)} + \Delta E_{jm} \qquad [4.72]$$

The analytical expression of the corrective term ΔE_{jm} is given in section 4.8.4.

4.5.4.1.4. Other cases

When the molecule's libration movement is performed around an equilibrium configuration $(\phi_0^{\mathrm{e}}, \theta_0^{\mathrm{e}})$ in which θ_0^e is neither close to 0 nor to $\pi/2$, the new angular coordinates used are $u_\theta = \sin(\theta_0 - \theta_0^{\mathrm{e}})$ and $u_\phi = \phi_0 - \phi_0^{\mathrm{e}}$. Introducing these coordinates into the kinetic rotation energy operator T_{rot} leads to an expression which highly depends on the value of θ_0^e. This is why we cannot give a detailed process for this situation, but are only indicating how to start.

4.5.5. *Active optical system*

This is composed of the vibration–orientation modes of the renormalized Hamiltonian of the trapped molecule:

$$\tilde{H}_{\mathrm{a}} = \tilde{H}_{\mathrm{vib}} + \tilde{H}_{\mathrm{orient}} \quad [4.73]$$

whose eigenenergies and eigenvectors are represented by $E_{\mathrm{v},jm}$ and $|\mathrm{v},jm\rangle$. By using the Born–Oppenheimer approximation, the eigenvectors of the optical system can be written in the form of tensor products $|\mathrm{v}\rangle \otimes |jm\rangle$ of the eigenvectors of vibration $|\mathrm{v}\rangle$ and the eigenvectors of orientation $|jm\rangle$. However, we must not forget that we must process the rovibrational movement of the non-rigid molecule. But, given the large difference between vibrational energy and orientation energy, it is possible to account for the influence of the vibration state on the orientation states by writing the eigenenergies and eigenvectors as: $E_{\mathrm{v},(jm)_{\mathrm{v}}}$ and $|jm\rangle_{\mathrm{v}}$. We must therefore replace the molecule's rotational constants B^e and centrifugal distortion constants D^e with B^{v} and D^{v} defined in equations [2.97] of Chapter 2.

4.5.6. *Translational modes*

The third term $\tilde{H}_{\mathrm{bath}}$ of equation [4.43] characterizes the thermal bath, which dissipates the energy coming from the molecule's optical modes. It is comprised of the translational vibration movements of the matrix atoms and the rigid molecule's center of mass evaluated at its equilibrium orientation and assumed to be a point defect:

$$\tilde{H}_{\mathrm{bath}} = H_{\mathrm{C}} + T_{\mathrm{trans}} + \hat{V}^{\mathrm{e}}_{\mathrm{MC}}(\vec{u}) \quad [4.74]$$

where H_{C} is the Hamiltonian of the perfect crystal defined in equation [4.13], T_{trans} is the translation kinetic Hamiltonian of the molecule's center of mass defined in equation [4.36] and $\hat{V}^{\mathrm{e}}_{\mathrm{MC}}(\vec{u})$ is, as we have already specified (equation [4.27]), the translation potential energy of the lattice containing the rigid molecule evaluated at its equilibrium orientation and assumed to be a point defect, from which the interaction energy between the substituted atom or atoms and the rest of the matrix outside the included molecule has been subtracted.

In the harmonic approximation of the translational vibrations of the "molecule–matrix" system, $\hat{V}^{\mathrm{e}}_{\mathrm{MC}}(\vec{u})$ can be written as:

$$\hat{V}^{\mathrm{e}}_{\mathrm{MC}}(\vec{u}) = \sum_{j}\left[\vec{\nabla}\hat{V}^{\mathrm{e}}_{\mathrm{MC}}\right]_{\vec{\xi}^{\mathrm{s}}_{j}}\vec{u}(j) + \frac{1}{2}\sum_{j,j'}\left[\vec{\nabla}\vec{\nabla}\hat{V}^{\mathrm{e}}_{\mathrm{MC}}\right]_{\vec{\xi}^{\mathrm{s}}_{j},\vec{\xi}^{\mathrm{s}}_{j'}}\vec{u}(j)\vec{u}(j') \qquad [4.75]$$

the linear terms $\left[\vec{\nabla}\hat{V}^{\mathrm{e}}_{\mathrm{MC}}\right]$ being equal to zero as they are calculated for the equilibrium positions, whereas the harmonic terms $\left[\vec{\nabla}\vec{\nabla}\hat{V}^{\mathrm{e}}_{\mathrm{MC}}\right]$ represent a force constant defect due to the presence of the molecule in the matrix.

The Hamiltonain $\tilde{H}_{\mathrm{bath}}$ can then be written in a compact form as:

$$\tilde{H}_{\mathrm{bath}} = \frac{1}{2}\sum_{j=0}\left(M_j\right)^{-1}P^2(j) + \frac{1}{2}\sum_{j,j'}\left(\varphi^{(2)}_{jj'}\right)\vec{u}(j)\vec{u}(j') \qquad [4.76]$$

where $\left(M_j\right)^{-1}$ is the masses' inverse matrix, which is diagonal and contains the mass defect due to the presence of the molecule, whereas $\left(\varphi^{(2)}_{jj'}\right)$ is the force constants matrix, which is non-diagonal and contains the force constant defect such that:

$$\begin{aligned}\left(M_j\right)^{-1} &= \left(M_j + (M_0 - M_j)\delta_{j,0}\right)^{-1}\\ \left(\varphi^{(2)}_{jj'}\right) &= \left(A(j,j') + (A(j,j';\vec{\xi}^{\mathrm{s}}) - A(j,j')) + \left[\vec{\nabla}\vec{\nabla}\hat{V}^{\mathrm{e}}_{\mathrm{MC}}\right]_{\vec{\xi}^{\mathrm{s}}_{j},\vec{\xi}^{\mathrm{s}}_{j'}}\right)\end{aligned} \qquad [4.77]$$

This force constants matrix is now independent of the optical variables $(q,\vec{\Omega}_0)$. We recall that $A(j,j')$ and $A(j,j';\vec{\xi}^{\mathrm{s}})$, respectively, represent the force constants of the perfect lattice and the distorted lattice by the fictitious presence of the molecule and in the absence of the substituted atom or atoms.

The eigenvalues and eigenvectors of the Hamiltonian $\tilde{H}_{\text{bath}}$ are then obtained by solving the dynamic equations:

$$\left(\mathrm{D} - \mathrm{I}\omega^2\right)\mathrm{u} = 0 \qquad [4.78]$$

which is of $3N$ dimensions if the molecule occupies a one substitutional site or of $3(N-1)$ dimensions if it occupies two substitutional sites in the lattice. In equation [4.78], D is the dynamic matrix of the "molecule–matrix" system, u represents the eigenvector associated with the eigenfrequency ω and I is the $3N$ or $3\ (N-1)$ dimensional identity matrix.

The trapped molecule presents a mass defect and a force constant defect. The dynamic matrix D can be separated into four blocks (submatrices): two diagonal blocks associated with the molecule and the atoms of the lattice, and two non-diagonal blocks containing the mass defects and the force constant defect. We have:

$$\mathrm{D} = \begin{pmatrix} \mathrm{D}^0 & \Delta\mathrm{D}^{0\mathrm{C}} \\ \Delta\mathrm{D}^{0\mathrm{C}} & \mathrm{D}^{\mathrm{C}} + \Delta\mathrm{D}^{\mathrm{C}} \end{pmatrix} \qquad [4.79]$$

where D^0 is the dynamic matrix associated with the translation movement of the molecule's center of mass in the static distorted lattice characterized by the variable $\vec{\xi}^{\mathrm{s}}$ and D^{C} is the dynamic matrix of the perfect crystal whose elements were defined in the expression [4.18]. The submatrices $\Delta\mathrm{D}^{\mathrm{C}}$ and $\Delta\mathrm{D}^{0\mathrm{C}}$, respectively, designate the difference in the dynamics of the crystal due to the static presence of the trapped molecule and the coupling between the movements of the molecule's center of mass and the atoms of the matrix. These two parts are generally small and are processed as perturbations.

Indeed, the resolution of equation [4.78] is mainly performed by using the Green function matrix G technique, which was dynamically developed for a monoatomic crystal containing a substitutional atomic defect with a mass and a forces constant, which are different from those of the substituted atom [MAR 65]. This is defined as the inverse of the dynamic matrix D, given by $\mathrm{G} = \mathrm{D}^{-1}$. It contains all the information on the movements of the bath and cannot be processed rigorously.

However, by considering some hypotheses and approximations, in particular, that:

1) only the binary interactions are considered;

2) the elements of the Green function matrix between two atoms *j* and *j*' or between an atom *j* and the molecule only depend on their relative distance (equation [4.25]).

the G matrix can be written as:

$$G \approx G^{0C}\left[I\text{-}(\Delta D^{C} + \Delta D^{0C})G^{0C}\right] \qquad [4.80]$$

Moreover, in this study, the bath only represents a possible dissipation channel of the energy of the optically active system, such that the perturbative corrections due to ΔD^{C} and ΔD^{0C} to the dynamic matrix of the doped crystal can be ignored. Thus, we can write:

$$G \approx G^{0C} = \begin{pmatrix} G^{0} & 0 \\ 0 & G^{C} \end{pmatrix} \qquad [4.81]$$

where G^{0} and G^{C} are, respectively, the Green function matrices associated with the translation dynamic of the molecule's center of mass and the perfect crystal.

The frequencies associated with the movement of the molecule's center of mass are obtained by diagonalizing the D^{0} matrix (3×3 matrix), whereas the frequencies associated with the movements of the atoms of the perfect crystal are contained in the density function of normalized phonons states $\omega^2 \nu(\omega^2)$ [FLI 62, MAN 68, MAN 71, MAR 65].

As we have already underlined previously in the case of the orientational modes, the translational modes depend, strictly speaking, on the molecule's internal vibration state.

4.5.7. *Optical modes – bath coupling*

Finally, the last term of equation [4.43] characterizes the coupling between the different movements of the molecule and the lattice. It now represents a perturbation term and can be expanded in the form:

$$\Delta\tilde{H}(q,\vec{\Omega}_0,\vec{u}) = \sum_j \left[\vec{\nabla} V_{\mathrm{MC}}(\vec{\Omega}_0)\right]_{\vec{\xi}_j^s} \vec{u}(j) + \frac{1}{2}\sum_{j,j'}\left[\vec{\nabla}\vec{\nabla} V_{\mathrm{MC}}(\vec{\Omega}_0)\right]_{\vec{\xi}_j^s,\vec{\xi}_{j'}^s} \vec{u}(j)\vec{u}(j') + \sum_j \left(\frac{\partial}{\partial q}\left[\vec{\nabla} V_{\mathrm{MC}}(\vec{\Omega}_0)\right]_{\vec{\xi}_j^s}\right)_{q=0} \vec{u}(j) q + \ldots \quad [4.82]$$

which can again be written as:

$$\Delta\tilde{H}(q,\vec{\Omega}_0,\vec{u}) = \Delta\tilde{H}_{1\mathrm{p}}(\vec{\Omega}_0,\vec{u}) + \Delta\tilde{H}_{2\mathrm{p}}(\vec{\Omega}_0,\vec{u}) + \Delta\tilde{H}'(q,\vec{\Omega}_0,\vec{u}) \quad [4.83]$$

The two first terms of this expression describe the dynamic coupling between the orientation movement of the rigid molecule and the phonon modes of the doped crystal (vibrations of the lattice and the molecule's center of mass). This coupling will be responsible for the energy's transfer mechanisms of orientation to the phonon modes (relaxation phase): (1) $\Delta\tilde{H}_{1\mathrm{p}}(\vec{\Omega}_0,\vec{u})$ for the one-phonon relaxation phase and (2) $\Delta\tilde{H}_{2\mathrm{p}}(\vec{\Omega}_0,\vec{u})$ for the two-phonon relaxation phase. This will enable a significant part of the shifts and broadening of the spectral lines to be determined, given that the modes concerned are in the range of low frequencies.

The third term of the expression [4.83] characterizes the vibration–orientation–phonons coupling responsible for the vibrational energy transfer mechanisms, high-frequency mode, toward the orientation modes and the low-frequency phonons. The effect of this coupling is clearly weaker and will be ignored for the rest of this chapter, which only focuses on the construction of the spectral profile.

4.6. Infrared absorption coefficient

4.6.1. *General expression*

The infrared absorption coefficient attached to a number N of optically active molecules per unit volume in a matrix and with no interaction between them is defined as the real part of the spectral density. In other words, Fourier transform of the time-dependent autocorrelation function $\Phi(t)$ is given as:

$$I(\omega) = \frac{4\pi N\omega}{3hc}\mathrm{Re}\int_0^\infty dt e^{i\omega t}\Phi(t)$$
$$\Phi(t) = \mathrm{Tr}\left[\rho(0)\mu_A(0)\mu_A(t)\right] \quad [4.84]$$

In this expression, μ_A is the dipole moment operator (vector) of the active molecule defined in the absolute frame (fixed frame), $\rho(0)$ is the initial canonical density matrix of the total system, ω is the frequency (expressed in wave number, cm^{-1}) and c is the speed of light. The symbol Tr (Trace) denotes the average on the initial conditions at the instant $t = 0$ and on all of the system's possible evolutions between the instant 0 and the instant t.

4.6.2. *Heisenberg representation*

In the Heisenberg representation, the time-dependent dipole moment operator $\mu_A(t)$ can be expressed in the form:

$$\mu_A(t) = e^{i\hbar^{-1}Ht}\mu_A(0)e^{-i\hbar^{-1}Ht} = e^{i\hbar^{-1}Lt}\mu_A(0) \quad [4.85]$$

where L is an operator defined in the Liouville space whose elements are the tensor products of the vectors and their conjugate vectors of the Hilbert space. It should be noted that in the Liouville space, the calculations are performed between the transitions coupling two Hilbert space levels and not between the Hilbert space levels. The Liouville operator is associated with the total Hamiltonian of the system by the commutator.

$$L\mu_A = [H, \mu_A] = (H\mu_A - \mu_A H) \quad [4.86]$$

It can be written as the sum of the three terms:

$$L = \tilde{L}_a + \tilde{L}_b + \Delta\tilde{L} \quad [4.87]$$

where the Liouville operators are, respectively, associated with the active optical system, i.e. the vibration–orientation modes of the molecule, the bath, i.e. the vibrations of the lattice and the molecule's center of mass, and with the coupling between the different movements, which are considered as a perturbation.

The adiabatic approximation allows the operator $e^{i\hbar^{-1}Lt}$ of the Liouville space to be written in the form:

$$e^{i\hbar^{-1}Lt} = U_L(t)e^{i\hbar^{-1}\tilde{L}_b t}e^{i\hbar^{-1}\tilde{L}_a t} \quad [4.88]$$

where $U_L(t)$ is the temporal evolution operator acting in the Liouville space and defined by:

$$U_L(t) = e^{i\hbar^{-1}\int_0^t \Delta\tilde{L}(t')dt'}$$

with [4.89]

$$\Delta\tilde{L}(t') = e^{i\hbar^{-1}\tilde{L}_b t'} e^{i\hbar^{-1}\tilde{L}_a t'} \Delta\tilde{L}(0)$$

and where E_i, E_f,... and $|i\rangle \equiv |v_i j_i m_i\rangle$, $|f\rangle \equiv |v_f, j_f m_f\rangle$,... are the eigenenergies and eigenvectors of the Hamiltonian of the optically active system $\tilde{H}_a$. The eigenelements of the associated Liouville operator are defined by $|i, f\rangle\rangle = |i\rangle \times \langle f|$ and $\omega_{if} = \hbar^{-1}(E_f - E_i)$, and they characterize the transition $i \to f$ between the initial state i and the final state f associated with the spectral line considered.

The autocorrelation function $\Phi(t)$ of the expression [4.84] cannot be calculated analytically (exactly). However, a semianalytic calculation is possible based on certain hypotheses and approximations, such as:

1) the permanent dipole moment of the molecule is much larger than the induced dipole moment resulting from its inclusion in the matrix, and does not depend on the modes of the bath;

2) because of the renormalization procedure of the total Hamiltonian, the coupling Hamiltonian $\Delta\tilde{H}$, which is initially small, remains as such throughout the time, a property that justifies the use of the initial *chaos hypothesis* (initial decoupling of the optical system and of the bath) allowing the initial canonical density matrix of the total system to be factorized as the product $\rho(0) = \rho_a(0)\rho_b(0)$ of the initial density matrices $\rho_a(0)$ and $\rho_b(0)$, which are respectively associated with the optically active system and with the bath and which are diagonal;

3) the spectral lines are assumed to be isolated from each other. This allows the "linked cluster theorem" to be applied, allowing only the diagonal matrix elements of $\langle U_L(t)\rangle_b$ to be calculated in the Liouville basis associated with the operator $\tilde{L}_a$, $\langle U_L(t)\rangle_b$ being the average on the bath states.

Under these conditions, the autocorrelation function $\Phi(t)$ can be written as:

$$\begin{aligned}\Phi(t) &= \mathrm{Tr}_{\mathrm{a}}\left[\rho_{\mathrm{a}}(0)\mu_A(0)\langle U_L(t)\rangle_{\mathrm{b}} e^{i\hbar^{-1}\tilde{L}_{\mathrm{a}}t}\mu_A(0)\right] \\ &= \sum_{i,f}(\rho_{\mathrm{a}}^i-\rho_{\mathrm{a}}^f)\langle f|\mu_A(0)|i\rangle\langle\langle if|\langle U_L(t)\rangle_{\mathrm{b}}|if\rangle\rangle e^{-i(\omega_{if}+\Delta\omega_{v_iv_f})t}\langle i|\mu_A(0)|f\rangle\end{aligned} \quad [4.90]$$

where ρ_{a}^i and ρ_{a}^f are, respectively, the populations, at the initial instant, of the initial i and final f optical states, and $\Delta\omega_{v_iv_f}$ is the shift of the vibrational frequency as discussed in section 5.3. It is evident that the information regarding the evolution of the form of spectral lines is contained in the matrix elements of the average of the evolution operator $\langle U_L(t)\rangle_{\mathrm{b}}$.

The matrix elements of the average of the evolution operator can be evaluated using the so-called "average cumulants expansion" technique up to the second order with respect to $\Delta\tilde{L}$ [ROB 76]:

$$\langle\langle if|\langle U_L(t)\rangle_{\mathrm{b}}|if\rangle\rangle = \exp\left\{\begin{aligned}&i\hbar^{-1}\int_0^t dt'\langle\langle if|\langle\Delta\tilde{L}(t')\rangle_{\mathrm{b}}|if\rangle\rangle - \\ &\hbar^{-2}\int_0^t dt'\int_0^{t'}dt''\left[\begin{aligned}&\langle\langle if|\langle\Delta\tilde{L}(t'')\Delta\tilde{L}(t')\rangle_{\mathrm{b}}|if\rangle\rangle - \\ &\langle\langle if|\langle\Delta\tilde{L}(t'')\rangle_{\mathrm{b}}|if\rangle\rangle\langle\langle if|\langle\Delta\tilde{L}(t')\rangle_{\mathrm{b}}|if\rangle\rangle\end{aligned}\right] \\ &+\ldots\ldots\end{aligned}\right\} \quad [4.91]$$

Moreover, due to the total Hamiltonian renormalization, the renormalized basis contains the permanent effect (stationary) of the "optical system–bath" interaction. Consequently, the collisional processes between the molecule and the bath can be considered as sudden, i.e. infinitely short in time. The *impact approximation* can then be used leading to a Lorentzian shape for the spectral lines.

Finally, by using the expressions [4.84], [4.90] and [4.91] and by replacing the matrix elements of the Liouville space with those of the Hilbert space, the absorption coefficient can be written in the form:

$$\begin{aligned}I(\omega) = \frac{4\pi N\omega}{3hc}\sum_{i,f}(\rho_{\mathrm{a}}^i-\rho_{\mathrm{a}}^f)\left|\langle i|\mu_A|f\rangle\right|^2 \mathrm{Re}\int_0^\infty dt\exp\left\{i(\omega-\omega_{if}-\Delta\omega_{v_iv_f})t\right\} \\ \exp\left\{-i\Delta^{(1)}t\right\}\exp\left\{-i\Delta^{(2)}t\right\}\exp\left\{-\Gamma^{(2)}t\right\}\end{aligned} \quad [4.92]$$

In this expression, $\Delta^{(1)}$ is the frequency shift of the line associated with the transition $i \to f$ due to the first-order term of the average cumulants expansion of the temporal evolution operator, whereas $\Delta^{(2)}$ and $\Gamma^{(2)}$ are the frequency shifts and broadening of this same line and are due to the second-order term of the expansion. They are given by the expressions:

$$\Delta^{(1)} = \hbar^{-1}\left[\left\langle f\middle|\left\langle \Delta\tilde{H}\right\rangle_{\text{b}}\middle|f\right\rangle - \left\langle i\middle|\left\langle \Delta\tilde{H}\right\rangle_{\text{b}}\middle|i\right\rangle\right] \quad [4.93]$$

$$\begin{aligned}\Delta^{(2)} &= \text{Im}\left[\hbar^{-2}\left(\Delta_i^{(2)} + \Delta_f^{(2)} + \Delta_{if}^{(2)}\right)\right]\\ \Gamma^{(2)} &= \text{Re}\left[\hbar^{-2}\left(\Delta_i^{(2)} + \Delta_f^{(2)} + \Delta_{if}^{(2)}\right)\right]\end{aligned} \quad [4.94]$$

in which

$$\Delta_i^{(2)} = \int_0^\infty d\tau \sum_{i'\neq i} e^{i(\omega_{i'}-\omega_i)\tau}\left\langle i\middle|\left\langle \Delta\tilde{H}(0)\right\rangle_{\text{b}}\middle|i'\right\rangle\left\langle i'\middle|\left\langle \Delta\tilde{H}(\tau)\right\rangle_{\text{b}}\middle|i\right\rangle \quad [4.95]$$

$$\Delta_f^{(2)} = \int_0^\infty d\tau \sum_{f'\neq f} e^{-i(\omega_{f'}-\omega_f)\tau}\left\langle f\middle|\left\langle \Delta\tilde{H}(\tau)\right\rangle_{\text{b}}\middle|f'\right\rangle\left\langle f'\middle|\left\langle \Delta\tilde{H}(0)\right\rangle_{\text{b}}\middle|f\right\rangle \quad [4.96]$$

and

$$\begin{aligned}\Delta_{if}^{(2)} = &\int_0^\infty d\tau\left\langle i\middle|\left\langle \Delta\tilde{H}(0) - \left\langle \Delta\tilde{H}(0)\right\rangle_{\text{b}}\middle|i\right\rangle\left\langle i\middle|\Delta\tilde{H}(\tau) - \left\langle \Delta\tilde{H}(0)\right\rangle_{\text{b}}\right\rangle_{\text{b}}\middle|i\right\rangle\\ &+\int_0^\infty d\tau\left\langle f\middle|\left\langle \Delta\tilde{H}(0) - \left\langle \Delta\tilde{H}(0)\right\rangle_{\text{b}}\middle|f\right\rangle\left\langle f\middle|\Delta\tilde{H}(\tau) - \left\langle \Delta\tilde{H}(0)\right\rangle_{\text{b}}\right\rangle_{\text{b}}\middle|f\right\rangle\\ &-\int_0^\infty d\tau\left\langle i\middle|\left\langle \Delta\tilde{H}(0) - \left\langle \Delta\tilde{H}(0)\right\rangle_{\text{b}}\middle|i\right\rangle\left\langle f\middle|\Delta\tilde{H}(\tau) - \left\langle \Delta\tilde{H}(0)\right\rangle_{\text{b}}\right\rangle_{\text{b}}\middle|f\right\rangle\\ &-\int_0^\infty d\tau\left\langle f\middle|\left\langle \Delta\tilde{H}(0) - \left\langle \Delta\tilde{H}(0)\right\rangle_{\text{b}}\middle|f\right\rangle\left\langle i\middle|\Delta\tilde{H}(\tau) - \left\langle \Delta\tilde{H}(0)\right\rangle_{\text{b}}\right\rangle_{\text{b}}\middle|i\right\rangle\end{aligned} \quad [4.97]$$

are given in terms of averages and correlation functions of the "orientation–bath" coupling Hamiltonian on the states of the bath.

The contribution $\Delta^{(1)}$ only characterizes a frequency shift of the line considered. More particularly, only the two-phonon coupling $\Delta\tilde{H}_{2\text{p}}(\vec{\Omega}_0,\vec{u})$ contributes to this shift, the average $\left\langle \Delta\tilde{H}_{1\text{p}}(\vec{\Omega}_0,\vec{u})\right\rangle_{\text{b}}$ of the one-phonon coupling being equal to zero in connection to the harmonic description of the translational modes.

The contributions $\Delta_i^{(2)}$ and $\Delta_f^{(2)}$ given in equations [4.95] and [4.96] characterize the inelastic collisional processes. They depend on the non-diagonal matrix elements describing the contamination of the initial and final optical states considered by the other optical states via the states of the bath.

On the other hand, the contribution $\Delta_{if}^{(2)}$ of equation [4.97] only characterizes the elastic collisions. For the initial and final optical states, they depend on the average on the bath states of the coupling Hamiltonian fluctuations around its average value.

4.6.3. *Averages and correlation functions*

The averages and correlation functions in expressions [4.93]–[4.97] can be determined by using the "orientation–bath" coupling terms of equation [4.81], corresponding to one phonon $\Delta\tilde{H}_{1p}(\vec{\Omega}_0,\vec{u})$ and to two phonons $\Delta\tilde{H}_{2p}(\vec{\Omega}_0,\vec{u})$.

These two terms can be expanded in terms of spherical harmonics as follows:

$$\Delta\tilde{H}_{1p}(\vec{\Omega}_0,\vec{u}) = \sum_{\alpha,j}\sum_{l,p}^{l\neq 0} A_\alpha^{lp}(\vec{\xi}_j^{\,s})Y_l^p(\theta_0,\phi_0)\times u_\alpha(j) \qquad [4.98]$$

$$\Delta\tilde{H}_{2p}(\vec{\Omega}_0,\vec{u}) = \sum_{\alpha',j'}\sum_{\alpha,j}\sum_{l,p}^{l\neq 0} A_{\alpha\alpha'}^{lp}(\vec{\xi}_j^{\,s},\vec{\xi}_{j'}^{\,s})Y_l^p(\theta_0,\phi_0)\times u_\alpha(j)u_{\alpha'}(j') \qquad [4.99]$$

where the coefficients A, respectively, represent the first and second derivatives of the interaction potential energy V_{MC}, between the molecule (j = 0) and the atoms of the matrix with respect to the translation coordinates along the directions $(\alpha = X,Y,Z)$.

The averages and correlation functions, on the bath states, connecting the dynamic displacement $u_\alpha(j)$ along the direction α at the instant 0 or τ to the dynamic displacement $u_{\alpha'}(j')$ following the direction α' at the instant 0, are generally expressed in terms of the Green function matrix elements $G_{\alpha\alpha'}$ of the molecule–matrix system defined in equation [4.81]. They are given by the expressions:

$$\langle u_\alpha(j,0)u_{\alpha'}(j',0)\rangle_b = -\frac{\hbar}{\pi}\int_0^{+\infty} d\omega \coth(\hbar\omega/2k_BT) \operatorname{Im} G_{\alpha\alpha'}(j,j',\omega^2 - i0) \quad [4.100]$$

$$\langle u_\alpha(j,\tau)u_{\alpha'}(j',0)\rangle_b = -\frac{\hbar}{\pi}\int_{-\infty}^{+\infty} d\omega \frac{\text{sign}(\omega)}{1\text{-}e^{-\hbar\omega/k_BT}} e^{-i\omega\tau} \operatorname{Im} G_{\alpha\alpha'}(j,j',\omega^2 - i0) \quad [4.101]$$

where $\operatorname{Im} G_{\alpha\alpha'}$ denotes the imaginary part of the elements $G_{\alpha\alpha'}$, k_B is the Boltzmann constant, T is the absolute temperature of the system and ω is the angular frequency of the perturbed crystal.

In section 4.5.6, based on some approximations, we were able to separate the Green function matrix G into two blocks: G^0 associated with the molecule's translation movement and G^C associated with the vibrations of the perfect lattice. Moreover, Table 4.4 indicates that the values of the matrix elements G^C are negligible following the different directions, and that these values decrease rapidly when the distance between atoms increases.

The above comments allow only the averages $\langle u_\alpha(j,0)u_\alpha(j,0)\rangle_b$ and "auto-correlation" functions $\langle u_\alpha(j,\tau)u_\alpha(j,0)\rangle_b$ for $(\alpha = X,Y,Z)$ to be taken into account in the calculations.

Thus, the elements of the Green function matrix are given by the relations:

$$1 + M_0\omega^2 G^0_{\alpha\alpha}(0,0;\omega^2 - i0) = -\frac{1}{3}\sum_{\alpha'}\omega^2_{\alpha'}\left[\text{PP}\left(\frac{1}{\omega^2 - \omega^2_{\alpha'}}\right) + i\pi\delta(\omega^2 - \omega^2_{\alpha'})\right] \quad [4.102]$$

for the molecule, and

$$1 + M_j\omega^2 G^C_{\alpha\alpha}(j,j;\omega^2 - i0) = -S(\omega^2) - i\pi\omega^2\nu(\omega^2) \quad [4.103]$$

with

$$S(\omega^2) = \text{p.v.}\int_0^\infty \frac{\omega'^2 \, \nu(\omega'^2)}{\omega^2 - \omega'^2} d\omega'^2 \quad [4.104]$$

for the atoms of the crystal; the symbols p.v. and δ designate respectively the Cauchy principal value of the integral and the Dirac function. As an example, Figure 4.4 illustrates the density of phonons $\omega^2\nu(\omega^2)$ and the function $S(\omega^2)$ of the perfect argon crystal.

Thus, we observe that the transfer of orientation energy is possible for the range of frequencies between 0 and ω_{max}, the cut-off frequency of the density of phonon states of the crystal (68 cm^{-1} in the case of the argon, for example), whereas it can only be achieved in resonance with the translational vibration frequencies of the molecule's center of mass or the localized mode. The existence of this mode depends on the mass defect and the force constant defect. As a matter of fact, when the molecule is much lighter than the atoms of the matrix, or when the force constant of the movement of the molecule's center of mass is clearly larger than that of the matrix's atoms, a localized mode of frequency $\omega_L \rangle \omega_{\text{max}}$ shows.

Finally, in the expressions [4.95]–[4.97], when the two-phonon coupling is introduced, the correlation functions are evaluated using the following Debye approximation:

$$\begin{aligned}\langle u_\alpha(j,\tau)u_{\alpha'}(j',\tau)u_\alpha(j,0)u_{\alpha'}(j',0)\rangle_b \approx{}& 2\langle u_\alpha(j,\tau)u_\alpha(j,0)\rangle_b \times \\ & \langle u_{\alpha'}(j',\tau)u_{\alpha'}(j',0)\rangle_b + \langle u_\alpha^2(j,0)\rangle_b \langle u_{\alpha'}^2(j',0)\rangle_b\end{aligned} \quad [4.105]$$

4.6.4. *Bar spectrum or Dirac spectrum*

This is an important step in the construction of the spectrum, because it informs us of the quality of the chosen basis and gives us an idea of the initial amplitude of the coupling between the optical system and the bath. Indeed, regarding the extent to which the bar spectrum enables the majority of the signals of the experimental spectrum to be reconstructed, we can admit that the coupling is and remains a small perturbation throughout the time evolution of the system as a whole.

After renormalizing the Hamiltonian of the total system, letting $U_L(t) = \mathbf{1}$ therefore means that the coupling $\Delta\tilde{H}(q,\vec{\Omega}_0,\vec{u})$ is assumed to be negligible. Equations [4.84] and [4.90] lead to the integrated absorption coefficients (bar lines) expressed as:

$$I^0(\omega) = \frac{8\pi^2 N\omega}{3hc}\sum_{i,f}(\rho_{\mathrm{a}}^i - \rho_{\mathrm{a}}^f)\left|\left\langle i\left|\left(\frac{d\vec{\mu}_A}{dq}\right)_{q=0} q\right|f\right\rangle\right|^2 \times\delta(\omega-\omega_{if}-\Delta\omega_{\mathrm{v}_i\mathrm{v}_f}) \qquad [4.106]$$

in the case of transitions implying the vibration–orientation states $|\mathrm{v}_i, j_i m_i\rangle \rightarrow |\mathrm{v}_f, j_f m_f\rangle$ (near infrared spectrum), and as:

$$I^0(\omega) = \frac{8\pi^2 N\omega}{3hc}\sum_{i,f}(\rho_{\mathrm{a}}^i - \rho_{\mathrm{a}}^f)\left|\left\langle i\left|\vec{\mu}_A^e\right|f\right\rangle\right|^2 \times\delta(\omega-\omega_{if}) \qquad [4.107]$$

in the case of transitions only implying the orientation states $|\mathrm{v}_i, j_i m_i\rangle \rightarrow |\mathrm{v}_i, j_f m_f\rangle$ (far infrared spectrum).

In the expressions [4.106] and [4.107], the population ρ_{a}^i (respectively, ρ_{a}^f) of the initial state (respectively, final state) of the optically active system is expressed as a function of its energy E_i (respectively, E_f) and the partition function Z, as:

$$\rho_{\mathrm{a}}^i = \frac{e^{-E_i/k_B T}}{Z} = \frac{e^{-E_i/k_B T}}{\sum_n e^{-E_n/k_B T}} \qquad [4.108]$$

and of the transitions elements of the dipole moment in the fixed frame as given in Chapter 2 (equations [2.92] and [2.100]).

4.6.5. *Spectral profile*

When the "orientation–bath" coupling is introduced, the shifts of the lines $\Delta^{(1)}$ and $\Delta^{(2)}$, and their homogenous width $\Gamma^{(2)}$ can be calculated. In the Lorentzian model hypothesis, the spectral profile of the trapped molecule can finally be written as:

$$I(\omega)=\frac{8\pi^2 N\omega}{3hc}\sum_{v_i,v_f}\left|\langle v_i|q|v_f\rangle\right|^2\sum_{\substack{j_i,m_i\\ j_f,m_f}}\frac{e^{-E_{v_i j_i m_i}/k_BT}-e^{-E_{v_f j_f m_f}/k_BT}}{Z}\left|\left\langle j_i m_i\left|\left(\frac{d\vec{\mu}_A}{dq}\right)_{q=0}\right|j_f m_f\right\rangle\right|^2$$
$$\times\frac{\Gamma^{(2)}}{(\omega-\omega_{v_i j_i m_i\to v_f j_f m_f}-\Delta\omega_{v_i v_f}-\Delta^{(1)}-\Delta^{(2)})^2+(\Gamma^{(2)})^2} \quad [4.109]$$

in the case of vibration–orientation transitions, and as:

$$I(\omega)=\frac{8\pi^2 N\omega}{3hc}\sum_{\substack{j_i,m_i\\ j_f,m_f}}\frac{e^{-E_{j_i m_i}/k_BT}-e^{-E_{j_f m_f}/k_BT}}{Z}\left|\left\langle j_i m_i\left|\vec{\mu}_A^e\right|j_f m_f\right\rangle\right|^2$$
$$\times\frac{\Gamma^{(2)}}{(\omega-\omega_{j_i m_i\to j_f m_f}-\Delta^{(1)}-\Delta^{(2)})^2+(\Gamma^{(2)})^2} \quad [4.110]$$

in the case of orientation transitions only.

As we have already specified in section 4.5.5, we need to recall that orientation states must be determined for each vibrational state.

4.7. Conclusion

A molecule trapped in a nanocage, either a substitutional site of a crystal (rare gas or nitrogen matrix) or an insertion site of clathrate, fullerene or zeolite, is subjected to interaction forces with atoms or molecules of the host system. A perturbation of the vibrational and rotational energy levels of the molecule, which is no longer isolated, then results from this situation. The frequencies of the vibrational transitions are shifted, whereas the rotation spectrum, as well as the translation of the molecule's center of mass, can be completely modified. The inclusion theoretical model or the Lakhlifi–Dahoo extended model described in this chapter enables the modified spectrum to be calculated using an approach similar to that described in Chapter 3 for calculating the shifts and the widths of the lines. The theoretical formalism is based on the density operator in the Liouville space, the short- and mid-range interaction through an atom–atom potential, the cumulants theorem and the Born–Oppenheimer approximation applied to separate the slow modes, associated with the degrees of freedom related to the phonons, to the translation of the center of mass and to the perturbed rotation from the vibrational modes. The widths of the lines can be calculated by relaxing the B.O. approximation to evaluate the effect of the otherwise ignored couplings. Furthermore, as the atoms are in given positions in a crystal, the Green functions method enables the distortion around the engaged molecule to be taken into account when calculating the modified spectrum.

4.8. Appendix

4.8.1. *Expression of the dispersion–repulsion contribution of the energy of truncated binary interaction in the fourth order*

$$V_{\mathrm{LJ}} = 4\sum_j \sum_{\mathrm{n}=6,12} (-1)^{\mathrm{n}/2} \sum_i \varepsilon_{ij} \left(\frac{\sigma_{ij}}{|\vec{\mathrm{r}}_{0j}|}\right)^{\mathrm{n}} \sum_{l_0, p_0} \frac{4\pi}{(2l_0+1)} \, {}^{\mathrm{n}}F_{p_0}^{l_0}(r_i, \beta_i) Y_{l_0}^{p_0}(\Theta_{0j}, \Phi_{0j}) Y_{l_0}^{-p_0}(\theta_0, \phi_0)$$

with

$${}^{\mathrm{n}}F_0^0(r_i, \beta_i) = \left\{1 + \frac{\mathrm{n(n\text{-}1)}}{6}\left(\frac{r_i}{|\vec{\mathrm{r}}_{0j}|}\right)^2 + \frac{\mathrm{(n\text{-}1)n(n+1)(n+2)}}{120}\left(\frac{r_i}{|\vec{\mathrm{r}}_{0j}|}\right)^4\right\}\sqrt{4\pi}\, Y_0^0(\beta_i, 0)$$

$${}^{\mathrm{n}}F_0^1(r_i, \beta_i) = \mathrm{n}\left(\frac{r_i}{|\vec{\mathrm{r}}_{0j}|}\right)\left\{1 + \frac{\mathrm{(n\text{-}1)(n+2)}}{10}\left(\frac{r_i}{|\vec{\mathrm{r}}_{0j}|}\right)^2\right\}\sqrt{\frac{4\pi}{3}}\, Y_1^0(\beta_i, 0)$$

$${}^{\mathrm{n}}F_0^2(r_i, \beta_i) = \mathrm{n}\left(\frac{r_i}{|\vec{\mathrm{r}}_{0j}|}\right)^2\left\{\frac{\mathrm{n}+2}{3} + \frac{\mathrm{(n\text{-}1)(3n^2+14n+32)}}{126}\left(\frac{r_i}{|\vec{\mathrm{r}}_{0j}|}\right)^2\right\}\sqrt{\frac{4\pi}{5}}\, Y_2^0(\beta_i, 0)$$

$${}^{\mathrm{n}}F_0^3(r_i, \beta_i) = \mathrm{n}\left(\frac{r_i}{|\vec{\mathrm{r}}_{0j}|}\right)^3 \frac{\mathrm{(n+2)(n+4)}}{15}\sqrt{\frac{4\pi}{7}}\, Y_3^0(\beta_i, 0)$$

$${}^{\mathrm{n}}F_0^4(r_i, \beta_i) = \mathrm{n}\left(\frac{r_i}{|\vec{\mathrm{r}}_{0j}|}\right)^4 \frac{\mathrm{(n^3+12n^2+44n+48)}}{105}\sqrt{\frac{4\pi}{9}}\, Y_4^0(\beta_i, 0)$$

We note that the coefficients of the spherical harmonics in the functions ${}^{\mathrm{n}}F_{p_0}^{l_0}(r_i, \beta_i)$ (with $p_0 \neq 0$) are identical to those of the spherical harmonics in ${}^{\mathrm{n}}F_0^{l_0}(r_i, \beta_i)$.

4.8.2. *Rotation matrix*

The unitary rotation matrix transforming the fixed frame $(G, \vec{X}, \vec{Y}, \vec{Z})$ to the mobile frame tied to the diatomic molecule (linear) $(G, \vec{x}, \vec{y}, \vec{z})$ is given by:

$$\mathbf{M}(\phi,\theta)=\begin{pmatrix} \cos\phi\cos\theta & \sin\phi\cos\theta & -\sin\theta \\ -\sin\phi & \cos\phi & 0 \\ \cos\phi\sin\theta & \sin\phi\sin\theta & \cos\theta \end{pmatrix}$$

4.8.3. *Eigenvalues correction of the orientation Hamiltonian*

$$\begin{aligned}\Delta E_{jM} = \frac{1}{32\alpha}&\left[\left(\frac{K_6}{4\alpha^2}-B^e\right)\left(7-6M^2\right)-2\left(1-M^2\right)\right]\left(2j-|M|+1\right)\\ &+\left[\frac{3}{8}\left(\frac{K_4}{8\alpha^2}-B^e\right)-1\right]\left(2j-|M|+1\right)^2\\ &+\frac{1}{32\alpha}\left[5\left(\frac{K_6}{4\alpha^2}-B^e\right)-6\right]\left(2j-|M|+1\right)^3+\dots\end{aligned}$$

4.8.4. *Eigenvalues correction of the orientation Hamiltonian*

$$\begin{aligned}\Delta E_{jm} = B^e&\left[-\frac{5}{8}-\frac{1}{2}\left(j+\frac{1}{2}\right)^2+\left(\frac{\beta_\phi}{\beta_\theta}\right)^2\left(j+\frac{1}{2}\right)\left(m+\frac{1}{2}\right)\right]\\ &+\frac{K_{22}}{\beta_\phi^2\beta_\theta^2}\left(j+\frac{1}{2}\right)\left(m+\frac{1}{2}\right)+\frac{3K_{04}}{2\beta_\theta^4}\left(j^2+j+\frac{1}{2}\right)\\ &+\frac{3K_{40}}{2\beta_\phi^4}\left(m^2+m+\frac{1}{2}\right)+\frac{3K_{06}}{8\beta_\theta^6}\left(8j^3+16j^2+22j+9\right)\\ &+\frac{3K_{60}}{8\beta_\phi^6}\left(8m^3+16m^2+22m+9\right)\\ &+\frac{3K_{24}}{2\beta_\phi^2\beta_\theta^4}\left(j^2+j+\frac{1}{2}\right)\left(m+\frac{1}{2}\right)\\ &+\frac{3K_{42}}{2\beta_\phi^4\beta_\theta^2}\left(j+\frac{1}{2}\right)\left(m^2+m+\frac{1}{2}\right)+\dots\end{aligned}$$

5

Applications to HCl, CO, O_2 and N_2

The theoretical models introduced are here applied to different types of diatomic molecules: the heteronuclear molecules, hydrogen chloride and carbon monoxide, that are characterized by permanent dipole moment and the homonuclear molecules, dioxygen and dinitrogen, which do not have a permanent dipole moment. The homonuclear molecules O_2 or N_2 are inactive in IR absorption spectroscopy but active in Raman spectroscopy. They can be observed and studied by a rotational Raman Lidar in the atmosphere, using the theories developed for isolated or nanoperturbed molecules in order to determine the temperature of the atmosphere, a fundamental parameter for meteorology and climatology. The absorption spectroscopy of heteronuclear molecules enables the characteristics of the environment where they are present, such as the graphite surface (CO), nanocage or substitution site in a solid matrix (HCl), to be determined by comparing the spectrum of the isolated molecule to that of the nanoperturbed molecule.

5.1. The HCl heteronuclear molecule isolated and trapped in a matrix

5.1.1. *Molecule in the gaseous phase*

The hydrogen chloride molecule HCl has a strong dipole moment μ. This is characterized by the strong intensity of its rotational absorption spectrum in the far infrared field, made up of a certain number of lines separated from each other by a wave number of about 21.2 cm^{-1}.

Moreover, its pure vibrational spectrum is composed of a very intense line located in the mid infrared at 2,886 cm^{-1} corresponding to the transition $v = 0 \rightarrow v = 1$ (fundamental), and a certain number of low-intensity lines located in the near infrared and corresponding to the transitions $v = 0 \rightarrow v = 2$, etc.

To interpret the rovibrational spectrum, it is indispensable to consider the coupled vibration–rotation movements in the framework of the anharmonic oscillator.

Thus, the eigenenergy associated with a vibration–rotation level (v, *J*) is given by equations [2.96], [2.97a] and [2.97b] from Chapter 2 and is expressed as:

$$E_{v,J} = E_v + B^v J(J+1) - D^v \left[J(J+1)\right]^2, \qquad [5.1]$$

where *Ev* is the eigenenergy of the vibration level v associated with the anharmonic oscillator, and B^v and $D^v \approx D^e$ are, respectively, the rotational and centrifugal distortion constants of the non-rigid rotator:

$$B^v = B^e - \alpha^e (v + \tfrac{1}{2}) \qquad [5.2]$$

However, as the numerical values of HCl's vibration anharmonic force constants are lacking, the expression of the energy E_v (equation [2.47]) is replaced by the following:

$$E_v = \left(v + \frac{1}{2}\right)\hbar\omega_0 - \left(v + \frac{1}{2}\right)^2 \hbar\omega_0 \chi^e, \qquad [5.3]$$

where ω_0 is the eigen angular frequency of the harmonic oscillator and the term χ^e characterizes the oscillator's anharmonicity.

Table 5.1 gives the values of the different constants associated with the molecule $H^{35}Cl$ in the gaseous phase [GRA 88].

r^e (Å)	μ^e (D)	B^e (cm^{-1})	$\hbar\omega_0$ (cm^{-1})
1.275	2.490	10.59	2,990
$r^e_{G\text{-}Cl}$ (Å)	D^e (cm^{-1})	α^e (cm^{-1})	$\hbar\omega_0\chi^e$ (cm^{-1})
0.035	0.00053	0.302	52

Table 5.1. *Characteristic constants of the molecule $H^{35}Cl$. $r^e_{G\text{-}Cl}$ represents the distance between the molecule's center of mass and the chlorine atom*

By substituting equations [5.2] and [5.3] in equation [5.1], the vibration–rotation eigenenergy can be written as:

$$E_{v,J} = \left(v+\frac{1}{2}\right)\hbar\omega_0 + B^e J(J+1) - \alpha^e(v+\tfrac{1}{2})J(J+1) - \left(v+\frac{1}{2}\right)^2 \hbar\omega_0\chi^e - D^e\left[J(J+1)\right]^2 \quad [5.4]$$

As an example, Table 5.2 shows the eigenenergies of some rovibrational levels, the first level ($v = 0, J = 0$) being set to 0.

Level (v , J)	$E_{V,J}$ (cm^{-1})	Level (v , J)	$E_{v,J}$ (cm^{-1})
$v = 0$, $J = 0$	0	$v = 1$, $J = 0$	2,886.00
$v = 0$, $J = 1$	20.88	$v = 1$, $J = 1$	2,906.88
$v = 0$, $J = 2$	62.61	$v = 1$, $J = 2$	2,948.61
$v = 0$, $J = 3$	125.19	$v = 1$, $J = 3$	3,007.57

Table 5.2. *Eigenenergies of some rovibrational levels of the HCl molecule in the gaseous phase*

5.1.2. *Molecule trapped in rare gas matrix*

The center of mass of the HCl molecule is almost coincident with the chlorine atom's center of mass, such that the HCl molecule is almost spherical and can substitute a rare gas atom in argon, krypton and xenon at very low temperatures (of the order of 20 K). The face-centered cubic character of the cage may still be assumed to be valid despite the molecule's inclusion. Moreover, it is reasonable to only consider the interactions between the molecule at the center of the inclusion site and the 12 nearest neighbor atoms, at their equilibrium position in the perfect crystal [KON 83].

Based on the adiabatic approximation, and after renormalizing the system's total Hamiltonian, the model developed by Kono and Lin [KON 83] has led to the evaluation of the transition energy for pure vibration between the fundamental level and the first excited level resulting from the inclusion of the HCl molecule in the argon matrix at: $E_1 - E_0 = 2{,}871\ \text{cm}^{-1}$. The corresponding value in the gaseous phase is $2{,}886\ \text{cm}^{-1}$.

Furthermore, when the translation movements of the molecule's center of mass and the matrix's atoms (rigid matrix) are disregarded, the study of the molecule's orientation potential energy shows that it performs a rotation movement that is hardly perturbed. The energy function $V_{\text{MC}}^{\text{e}}(\phi_0,\theta_0)$ can be expanded in terms of spherical harmonics as (equation [4.51]):

$$V_{\text{MC}}^{\text{e}}(\phi_0,\theta_0)=\sum_{l_0}\sum_{p_0=-l_0}^{+l_0} A_{p_0}^{l_0}\times Y_{l_0}^{p_0}(\theta_0,\phi_0) \qquad [5.5]$$

However, because the molecule is found in a site of octahedral symmetry O_h, the values of the coefficients $A_{p_0}^{l_0}$ are only non-zero for the values of $l_0 = 2k + 4$ with $k = 0, 1, 2$, etc. and $p_0 = \pm 4k'$ with $k' = 0, 1, 2$, etc.

In the case of HCl in an argon matrix, the expansion performed by Kono and Lin shows that the first three terms corresponding to the coefficients $A_0^4 = -24.8\ \text{cm}^{-1}$ and $A_{-4}^4 = A_{+4}^4 = -14.8\ \text{cm}^{-1}$ are sufficient (higher order terms are negligible) for representing the rotation energy surface with a good approximation. Thus, the energy difference between the pure rotational levels $J = 0$ and $J = 1$ is evaluated as: $E_1 - E_0 \approx 21.0\ \text{cm}^{-1}$, which is $0.1\ \text{cm}^{-1}$ higher than that of the gaseous phase. This is a very small displacement that confirms the quasi-free rotation movement of HCl in the argon matrix.

This result is largely sufficient for the molecule to be considered as a point defect and to determine the frequency of the vibration movement (translation) of its center of mass as a function of the mass and the force constants defects with respect to the perfect crystal. In the case of HCl in the argon matrix, the mass defect is quite small, evaluated as $M_j - M_0 = 3.49$ g/mole, whereas the force constant defect is large, given by:

$\left(\frac{\partial^2 V^{\mathrm{e}}_{\mathrm{MC}}}{\partial u^2(0)}\right)-\left(\frac{\partial^2 V^{\mathrm{e}}_{\mathrm{C}}}{\partial u^2(j)}\right)=3252-2308=944\ \ \mathrm{cm}^{-1}.\mathrm{Å}^{-2}$. The frequency of the vibration modes of the molecule's center of mass is given by the equation:

$$1-\rho(\omega^2)S(\omega^2)=0 \tag{5.6}$$

where

$$\rho(\omega^2)=\left(-1+\frac{M_j}{M_0}\right)+\frac{2\omega^2}{\omega^2_{\max}}\left(1-\left(\frac{\partial^2 V^{\mathrm{e}}_{\mathrm{C}}}{\partial u^2(j)}\right)\Big/\left(\frac{\partial^2 V^{\mathrm{e}}_{\mathrm{MC}}}{\partial u^2(0)}\right)\right) \tag{5.7}$$

The solution to equation [5.6] then leads to the frequency $\omega_{\mathrm{L}}=73\ \mathrm{cm}^{-1}$, which thus corresponds well to a localized mode because it is greater than the perfect crystal's cut-off frequency $\omega_{\max}=68\ \mathrm{cm}^{-1}$.

Finally, when the rotation–translation coupling is taken into account, the displacement of the transition frequency between the rotational levels $J = 0$ and $J = 1$ is determined as –3.4 cm^{-1} with respect to those of the molecule in the gaseous phase. Moreover, a part of this displacement, –1.6 cm^{-1}, is due to the rotation-lattice's phonons coupling, and the other part, –1.8 cm^{-1}, is due to the rotation-molecule's localized mode coupling.

5.2. Lidar probing of terrestrial homonuclear molecules N_2 and O_2

The Lidar (light detection and ranging), which is the equivalent of a radar in the optical field, is a technique that consists of using a laser as a light source for illuminating a cloud and then recording the signals backscattered by the molecules present in the cloud to determine the characteristics, composition, temperature or pressure [HAU 80]. As the temperature is a fundamental parameter for meteorology and climatology, a rotational Raman Lidar can be used for monitoring weather and climate from ground, air or space devices. The vertical profile of the ground temperature in the mid stratosphere [KEC 90] can thus be determined based on the theoretical models described in Chapters 4 for analyzing the signals backscattered by the molecules of O_2 or N_2 present in the earth's atmosphere. The operator that induces the transition between the vibration–rotation levels (Chapter 1) is a rank 2 tensor and thus the selection rule is given by $\Delta J = \pm\, 2$ (Stokes and anti-Stokes) for the inelastic diffusion and $\Delta J = 0$ for the elastic diffusion (Rayleigh) (Figure 5.1).

The rotational Raman spectrum of a diatomic molecule (N_2 or O_2) thus comprises the Stokes branches ($J = 0,1,2$, etc.) and anti-Stokes branches ($J = 2,3$, etc.), each branch being made up of a series of lines with an intensity $I(J)$, which varies according to the temperature such as:

$$I(J) \approx g_n S_J \exp(-E(J)/kT) \qquad [5.8]$$

with (S_J(S) = $(J+1)(J+2)/(2J+3)$ and S_J(AS) = $(J-1)(J)/(2J-1)$ and g_n is such that:

$$g_n = \begin{cases} 6 \text{ for even } J \\ 3 \text{ for odd } J \end{cases} \quad \text{for } N_2 \qquad [5.9a]$$

and

$$g_n = \begin{cases} 0 \text{ for even } J \\ 1 \text{ for odd } J \end{cases} \quad \text{for } O_2 \qquad [5.9b]$$

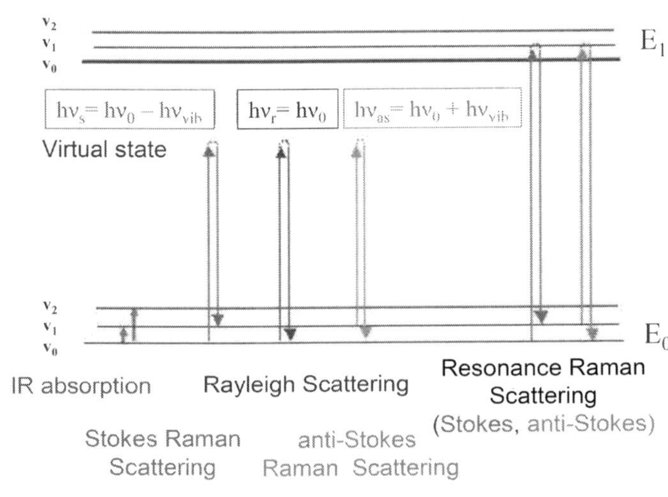

Figure 5.1. *Light–matter interaction: IR spectroscopy ($\Delta J = \pm 1$), Rayleigh ($\Delta J = 0$) and Raman ($\Delta J = \pm 2$). For a color version of this figure, see www.iste.co.uk/dahoo/infrared.zip*

The following program written for MATLAB allows one to calculate the intensities of the Stokes (anti-Stokes) lines for N_2. It can easily be optimized and adapted for O_2 by modifying the molecular parameters *b0*, *d0*, s_j and g_n, or for other diatomic molecules. The molecular parameters and the formulas for N_2 (and O_2) are given in [BEH 02, BUT 71, CLA 75][1].

```
clc
clear all
n = 1;                              % component n of the studied medium
for j = 0:20
kt = 0.027 × 8065;                  % (cm^-1)
b0 = 1.99;                          % Rotational constant (cm^-1)
d0 = 5.48 × 10^-6;                  % Centrifugal distortion constant (cm^-1)
for js=0 :1
if js == 0
   delta = 2 × b0 × ((2 × j) – 1);      % for anti-Stokes
   sj = ((j – 1) × j)/((2 × j) – 1);    % for anti-Stokes
else
   delta = –2 × b0 × ((2 × j) + 3);     % for Stokes
   sj = ((j + 1) × (j + 2))/((2 × j) + 3);   % for Stokes
end
if mod(j,2) == 0
  gn = 6;
else
  gn = 3;
end
Ej = b0 × j × (j + 1) – d0 × (j × (j + 1))^2;
i = gn × sj × exp(–Ej/kt);
b(n,1) = delta;
b(n,2) = i;
n = n + 1;
end
bar(b(:,1),b(:,2))
ylabel('Intensity');
xlabel('Energy (cm^-1)');
title('Raman Effect');
```

The measurement of the ratio between two opposite lines (Stokes and anti-Stokes) with a temperature dependency (Figure 5.2) enables the atmospheric temperature to be determined down to the surface without any hypothesis on the atmospheric transmission [HAU 16].

1 For a color version of this code, see www.iste.co.uk/dahoo/infrared.zip.

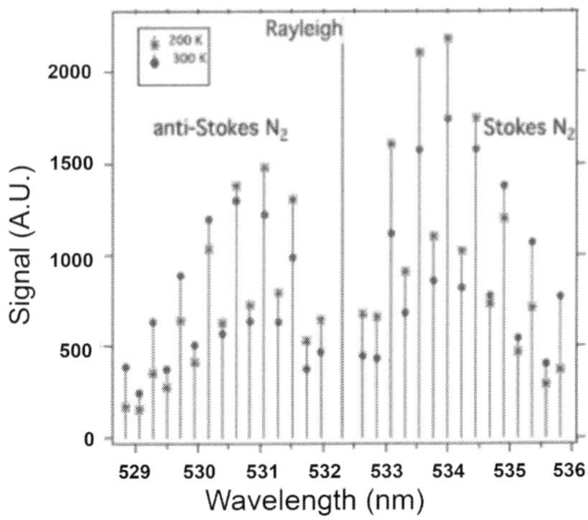

Figure 5.2. *Simulated N_2 Raman signal for 30 km of altitude, 100 m of thickness and 1 h of signal integration for two temperatures: 200 K and 300 K (532 nm laser). For a color version of this figure, see www.iste.co.uk/dahoo/infrared.zip*

Applying Robert and Bonamy's theory [ROB 79] described in Chapter 3 enables the time dependence of the line widths to be determined using:

$$\frac{\gamma^0(T)}{\gamma^0(T_0)} = \left(\frac{T_0}{T}\right)^n \qquad [5.10]$$

where $\gamma^0(T)$ is the broadening parameter (the half-width to half-maximum (HWHM) per unit pressure) at the temperature T (varying as T^{-n}) and T_0 is the ambient temperature. Figure 5.3 gives an example of the application of the theory to the calculation of the carbon dioxide molecule's line widths (CO_2) perturbed by the dinitrogen N_2 at the ambient temperature of 296 K and a lower temperature of 198 K [DAH 88]. The simulation gives values that are comparable to the experimental ones and enables the parameter n to be determined. In the case shown in Figure 5.3, the value of n is evaluated at 0.72, a relative error of 3% on the broadening parameter leading to an error of 0.13 on n.

Thus, in the simulated signal (Figure 5.2), one can include another parameter, in this case the pressure, to characterize the probed medium.

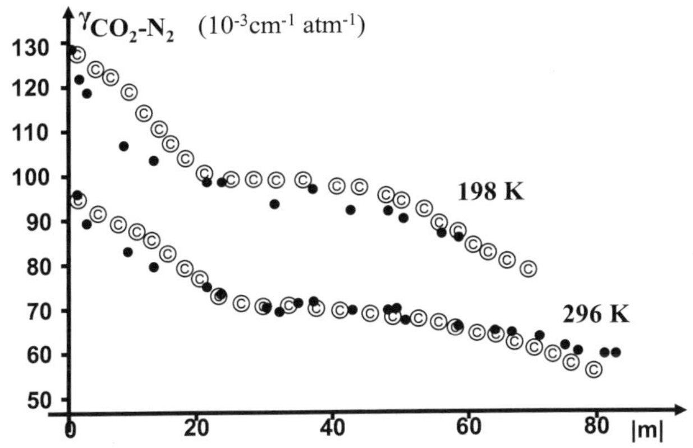

Figure 5.3. *Broadening parameters of $^{12}C^{16}O_2$ by N_2 at 296 K and 198 K [DAH 88]: • experimental values and © values calculated by the RB model. The quadrupole momenta are as follows: $Q_{N2}= -1.3 \times 10^{-26}$ esu cgs and $Q_{CO2} = -3.85 \times 10^{-26}$ esu cgs; the atom–atom parameters are from [MUR 80] and [MUR 81] with a non-rectilinear trajectory*

5.3. The heteronuclear molecule CO trapped in a matrix and absorbed on graphite substrate (1000) at a low temperature

5.3.1. *Molecule trapped in a rare gas matrix*

The carbon monoxide CO molecule was studied in matrices by Dubost [DUB 75, DUB 76]. The molecule's fundamental vibration mode is centered at 2,143.2 cm^{-1}. In the argon matrix, the CO molecule is trapped in a substitution site with a diameter of 0.376 nm and which is thus slightly deformed because the molecule's bond length is 0.113 nm, while the Van der Waals radii of the atoms O and C are, respectively, 0.14 and 0.165 nm. In an argon matrix, the vibrational frequency is centered at 2,138.64 cm^{-1}. The work presented in [DUB 75] shows that the broadening of the spectral line of the CO molecule depends on the temperature and that the molecule does not rotate in the nanocage, performing small oscillations or librations around well-defined given directions. The coexistence of two conformations of the stabilized CO–CO_2 complex in the argon matrix [RAD 95], that reversibly transform with the temperature variation, has been interpreted on the basis of three possible positions of the CO with respect to only a single one for CO_2 [RAD 96]. This interpretation has been corroborated by a model based on the methods of quantum chemistry [LAN 99]. These results are consistent with those obtained on the equilibrium configuration of the CO molecule in matrices.

Figure 5.4 represents the iso-surfaces of the interaction potential energy of the CO molecule in a xenon matrix for a face-centered cubic structure. The calculations (equation [4.1] of Chapter 4) show that the nanocage in which the molecule is trapped only consists of one single substitution site. The iso-surfaces calculated by rotating the CO molecule about the C_4 axis of symmetry of the fcc structure of the xenon show that CO is quasi-free in the nanocage, with a barrier to the rotation of 25 cm^{-1}. These results show that in the case of CO in a rare gas matrix, the IR absorption spectrum will not be a vibration–rotation spectrum comprising a P and R branch constituted by rotation lines like for HCl but a Q branch tied to the single vibrational degree of freedom with a librational or perturbed rotational structure. This prediction is consistent with the experimental results presented in [DUB 75, LAN 99], namely the observation of a single IR absorption line for the fundamental vibrational transition in the different rare gas matrices.

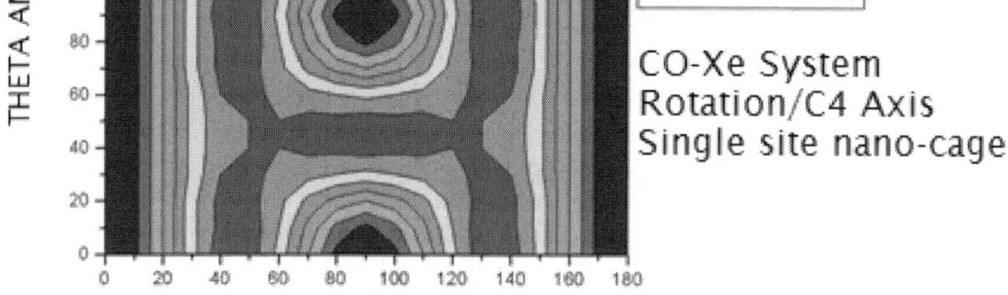

Figure 5.4. *Configuration energy (cm^{-1}) of the solid molecule–matrix system: CO/Xe; barrier at 25 cm^{-1} for rotation. For a color version of this figure, see www.iste.co.uk/dahoo/infrared.zip*

Table 5.3 shows the values of the different characteristic parameters of the CO molecule in the gaseous phase, defined in the frame $(G,\vec{x},\vec{y},\vec{z})$ tied to the molecule.

r^e (Å)		μ^e (D)		B^e (cm^{-1})	$\hbar\omega_0$ (cm^{-1})
1.144		0.112		1.93	2,169.8
α^e_{zz} ($Å^3$)	$\alpha^e_{xx} = \alpha^e_{yy}$	Θ^e_{zz} (DÅ)	$\Theta^e_{xx} = \Theta^e_{yy}$	α^e (cm^{-1})	$\hbar\omega_0\chi^e$ (cm^{-1})
2.600	1.625	–2.50	1.25	0.0175	13.29

Table 5.3. *Values of the CO molecule's characteristic parameters (non-negligible) in the gaseous phase*

5.3.2. *Molecule adsorbed on the graphite substrate*

The graphite crystal is often used to model the interstellar grains of dust in order to study the physical–chemical behavior of atomic or molecular species adsorbed on the surface of these grains.

Graphite has a hexagonal structure. It is well split in the cleavage direction (1000) offering a surface (substrate) that is sufficiently chemically inert and hardly deformed (relaxed). The crystallographic planes are parallel, separated by a distance L = 3.357 Å and are alternatively offset from each other with a periodicity of 2.

To each crystallographic plane p is associated a primitive cell, with a parameter a = 2.460 Å, containing two carbon atoms (s = 1, 2) and primitive translation vectors $\vec{a}_1$ and $\vec{a}_2$ defined in the fixed frame $(O,\vec{X},\vec{Y},\vec{Z})$, connected to the substrate, as:

$$\vec{a}_1 = a \times \vec{X}$$
$$\vec{a}_2 = \frac{a}{2} \times \vec{X} + \frac{a\sqrt{3}}{2} \times \vec{Y} \qquad [5.11]$$

Figure 5.3 shows the geometrical characteristics of the adsorption of a CO molecule on the graphite substrate as well as the definition of the primitive cell and its primitive vectors.

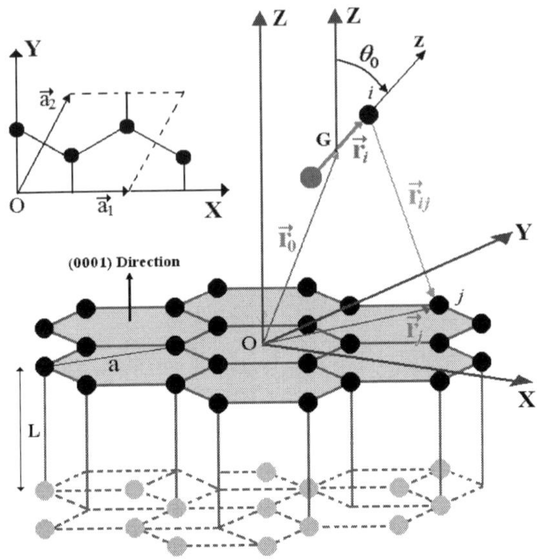

Figure 5.5. *Geometrical characteristics of a molecule's adsorption on the graphite substrate. The cell and the primitive vectors are defined on the top left. For a color version of this figure, see www.iste.co.uk/dahoo/infrared.zip*

Without going into detail, a carbon atom j of a given plane is described by four numbers: l_1, l_2, two relative integers $s = 1$ or 2 and p the plane number, $p = 1$ being the plane at the surface. The position vector $\vec{r}_j$ can then be written in the fixed frame as:

$$\vec{r}_j = \frac{a}{2}\left(2l_1 + l_2 + s - \text{mod}(\tfrac{p-1}{2})\right)\times\vec{X} + \frac{a}{2\sqrt{3}}\left(3l_2 + s - \text{mod}(\tfrac{p-1}{2})\right)\times\vec{Y} \\ -(p-1)L\times\vec{Z} \qquad [5.12]$$

in which the term mod $(\frac{p-1}{2})$ means "the remainders of the division of (p – 1) by 2".

Furthermore, the covalent bonds between the carbon atoms of the same plane give rise to aspherical atomic charge distributions, which generate an electric field outside the substrate. This phenomenon was modeled by placing a quadrupole moment Θ^c on each atom in the (1000) direction and whose value is estimated to be equal to 1 DÅ [VER 92].

On the other hand, the conduction or transport property per graphite plane induces an anisotropy of the dipolar polarizability of the carbon atoms whose matrix elements in the fixed frame are given by: $\alpha^{c}_{ZZ} = 0.413$ Å^3 and $\alpha^{c}_{XX} = \alpha^{c}_{YY} = 1.444$ Å^3.

5.3.3. *Molecule–graphite interaction energy*

The interaction potential energy V_{MS} between the molecule and the substrate is the sum of pairwise interaction potentials of the molecule–atom of graphite Gr, of the following contributions: dispersion–repulsion Lennard–Jones V_{LJ} type, induction V_I and electrostatic V_E.

5.3.3.1. *Dispersion–repulsion contribution*

In order to take the anisotropy phenomenon described above into account, Carlos and Cole [CAR 80] proposed a modified form of the atom–atom potential energy as:

$$V_{LJ} = \sum_i \sum_j 4\varepsilon_{ij} \left\{ \begin{array}{l} \left(\dfrac{\sigma_{ij}}{|\vec{r}_{ij}|}\right)^{12} \left[1+\gamma_R\left(1-\dfrac{6}{5}\cos^2\theta_{ij}\right)\right] \\ \qquad -\left(\dfrac{\sigma_{ij}}{|\vec{r}_{ij}|}\right)^{6} \left[1+\gamma_A\left(1-\dfrac{3}{2}\cos^2\theta_{ij}\right)\right] \end{array} \right\} \quad [5.13]$$

where γ_R and γ_A are coefficients associated with the anisotropy and θ_{ij} is the angle between the $\vec{Z}$ axis and the vector $\vec{r}_{ij}$ binding the atom i of the molecule and the atom j of graphite, which can be written as:

$$\vec{r}_{ij} = \vec{r}_j - \vec{r}_0 - \vec{r}_i \quad [5.14]$$

as a function of the position vectors of the carbon atom j and the molecule's center of mass in the fixed frame, and the position vector of the atom i of the molecule in the frame tied to the latter. This vector then depends on the molecule's internal vibration and orientation coordinates.

The values of the atom–atom parameters of the expression [5.13] are given in Table 5.4.

Atom–Gr	ε (cm^{-1})	σ (Å)	γ_A	γ_R
C–Gr	24.04	3.305	0.4	–1.05
O–Gr	27.82	3.141	0.4	–1.05

Table 5.4. *Atom–atom parameters of the Lennard–Jones potential and atom–atom parameters associated with the anisotropy of the Gr atoms*

5.3.3.2. *Electrostatic and induction contribution*

These contributions were introduced in Chapter 4. They describe the interaction between the permanent multipole moments of the molecule-Gr atom partners, and the interaction between these permanent moments and the electric field that they generate at the position of each partner:

$$V_{\mathrm{E}} = -\sum_j \vec{\mu}_0.\vec{E}_{j\to 0} - \frac{1}{3}\sum_j \vec{\vec{\Theta}}_0 : \vec{\nabla}\vec{E}_{j\to 0} \qquad [5.15]$$

$$V_{\mathrm{I}} = -\frac{1}{2}\vec{E}_{0\to j}.\vec{\vec{\alpha}}_j^{\mathrm{c}}.\vec{E}_{0\to j} - \frac{1}{2}\vec{E}_{j\to 0}.\vec{\vec{\alpha}}_0.\vec{E}_{j\to 0} \qquad [5.16]$$

5.3.4. *Adsorption observables at a low temperature*

The substrate used in our study is a crystal composed of 10 planes of 35 Å radius each in order to prevent the side effects that the molecule may experience.

As the molecule and the substrate are assumed to be rigid, two observables characterizing the adsorption can be determined: the adsorption energy and the molecule's diffusion constant on the surface.

First, the potential energy surface $V_{\mathrm{MS}}(X_0, Y_0)$ connecting the molecule's lateral movement is determined by minimizing $V_{\mathrm{MS}}(\vec{r}_0, \vec{\Omega}_0)$ with respect to the distance Z_0 between the molecule's center of mass and the surface, and the angular coordinates (ϕ_0, θ_0). The equilibrium molecule's configuration and site are associated with the energy's minimum value:

$V_{\mathrm{MS}}^{\mathrm{min}} = -960$ cm–1 for $(X_0^e, Y_0^e) = (0,0)$, $Z_0^e = 3.08$ Å and $\theta_0^e \approx 95°$, ϕ_0 taking any value. These values correspond to a nearly flat position of the molecule above the hexagon center.

On the other hand, the energy's maximum value $V_{\mathrm{MS}}^{\mathrm{max}} = -863\ \mathrm{cm}^{-1}$ is obtained when the molecule is found above a carbon atom in the flat position at a distance of 3.24 Å from the surface. Finally, a value of –870 cm^{-1} is obtained when the molecule is found to be in the flat position above the middle of an edge (C–C bond) of the hexagon perpendicular to the latter, and at a distance of 3.20 Å.

From these potential energy calculations, one can note that the dispersion–repulsion contribution represents about 95% of the total energy and that the surface plane (p = 1) contributes to about 85% of this total energy.

The potential energy surface $V_{\mathrm{MS}}(X_0, Y_0)$ is shown in Figure 5.4, as well as the diffusion path (the valley) between two equilibrium configurations. The diffusion energy barrier is equal to 90 cm^{-1}.

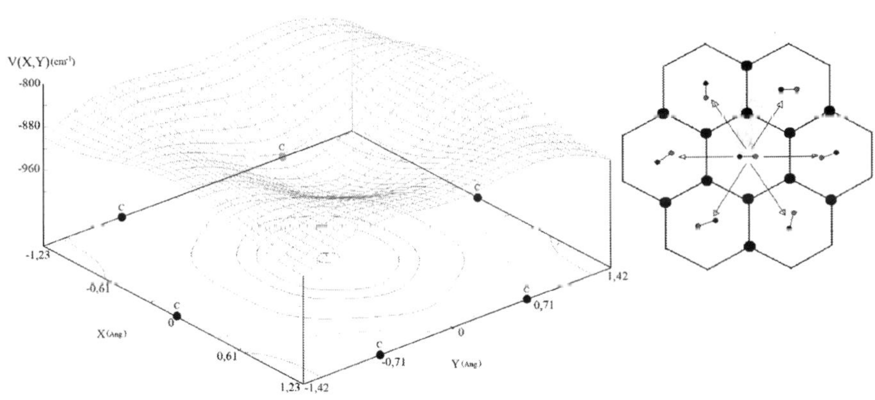

Figure 5.6. *Potential energy surface of the CO molecule's lateral movement on the graphite surface (left) and diffusion path from an equilibrium position to another (right). For a color version of this figure, see www.iste.co.uk/dahoo/infrared.zip*

The study of the external movements of the CO molecule shows that the vibrations of its center of mass are characterized by:

– large amplitudes in X_0 and Y_0;

– small amplitudes in Z_0;

and

– small amplitude librations in θ_0;

– a quasi-free movement in ϕ_0.

5.3.4.1. *Adsorption energy*

The adsorption energy E_a of an adsorbed molecule on a surface at a temperature T is defined as the energy necessary for maintaining this molecule, of thermal energy $(\frac{n}{2}+1)k_B T$, at its equilibrium configuration on this surface. It is given approximately by:

$$E_\text{a} \approx \left(\frac{n}{2}+1\right)k_B T - V_\text{MS}^\text{min} - \sum_{s=1}^{n}\frac{\hbar\omega_s}{2}\coth\left(\frac{\hbar\omega_s}{2k_B T}\right), \qquad [5.17]$$

where n is the number of the molecule's external degrees of freedom (5 for a diatomic molecule) and ω_s (s = 1 at n) are the frequencies associated with the translation movements and the molecule's orientation around its equilibrium configuration. The frequency values ω_s and the adsorption energy E_a according to the temperature are given in Table 5.5.

ω_X	ω_Y	ω_Z	ω_θ	ω_ϕ	E_a(10K)		E_a(100K)
26.6	26.6	62.1	76.6	~0	888	920	920

Table 5.5. *Frequencies (in cm^{-1}) of the translation and orientation movements of the adsorbed CO on the graphite surface and the adsorption energy (in cm^{-1}) as a function of the temperature*

5.3.4.2. *Diffusion constant*

Let $\Delta V_\text{MS}(\tau)$ be the variation of the potential energy along the diffusion path between the two adjacent equilibrium sites of respective positions τ_1 and τ_2, as shown in Figure 5.7. The set of values of the diffusion coordinate is in fact the set of the coordinate points (X_0, Y_0) with respect to the fixed frame.

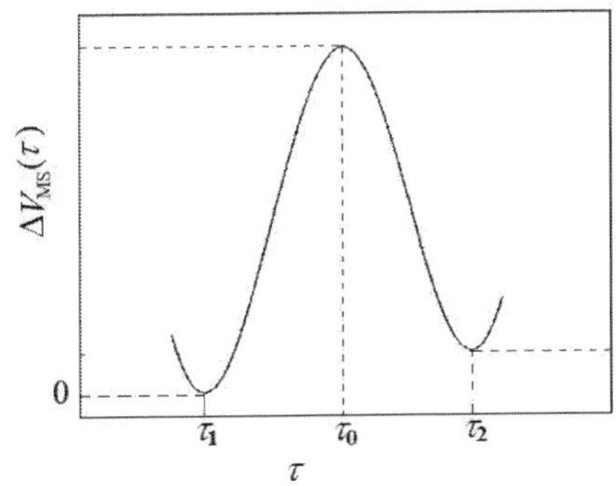

Figure 5.7. *Diffusion path between the two equilibrium sites. In the case of CO on the graphite surface, the two sites are equivalent*

By using the classical transition theory between states [VOT 84], the frequency of diffusion jumps above the maximum position τ_0 is defined by:

$$K(T) \approx \left(\frac{k_B T}{2\pi M}\right)^{1/2} \frac{\exp[-\Delta V_{MS}(\tau_0) / k_B T]}{\int_{\tau_1}^{\tau_2} d\tau \exp[-\Delta V_{MS}(\tau) / k_B T]}, \qquad [5.18]$$

where M is the molecule's mass and $\Delta V_{MS}(\tau_0)$ is the energy's value at the maximum of the diffusion path. The diffusion constant is then given by:

$$D(T) = (\tau_2 - \tau_1)^2 K(T) \qquad [5.19]$$

Table 5.6 shows the distance values between the two sites, the height of the energy barrier and the diffusion constants at different temperatures.

$(\tau_2 - \tau_1)$ (Å)	$\Delta V_{MS}(\tau_0)$ (cm^{-1})	D(10K) (cm^2/s)	D(40K) (cm^2/s)	D(100K) (cm^2/s)
2.460	90	3.3×10^{-10}	1.0×10^{-5}	7.3×10^{-5}

Table 5.6. *Diffusion characteristics of CO on the graphite surface*

5.4. Conclusion

In this chapter, numerical calculations have been given for some examples to illustrate the application of the theoretical models described in the previous chapters. The simulation based on the theoretical developments helps in better defining observation methods and also in the interpretation of the observations themselves.

Bibliography

[ACH 96] ACHIMSKY L., Etude Cinétique de la thermo-oxydation du polypropylène, PhD Thesis, ENSAM, Paris, 1996.

[ADA 41] ADAMS W.S., "Some Results with the COUDÉ Spectrograph of the Mount Wilson Observatory", *Astrophysical Journal*, vol. 93, pp. 11–23, 1941.

[ALI 85] ALIEV D., WATSON J.K.G., "Higher-Order Effects in the Vibration-Rotation Spectra of Semi-rigid Molecules", in NARAHARI RAO K. (ed.), *Molecular Spectroscopy: Modern Reseach*, vol. 3, Academic Press, Orlando, 1985.

[AMA 57a] AMAT G., GOLDSMITH M., NIELSEN H.H., "Higher Order Rotation-Vibration Energies of Polyatomic Molecules II", *J. Chem. Phys.*, vol. 27, pp. 838–844, 1957.

[AMA 57b] AMAT G., NIELSEN H.H., "Higher Order Rotation-Vibration Energies of Polyatomic Molecules III", *J. Chem. Phys.*, vol. 27, pp. 845–850, 1957.

[AMA 58] AMAT G., NIELSEN H.H., "Higher Order Rotation-Vibration Energies of Polyatomic Molecules IV", *J. Chem. Phys.*, vol. 29, pp. 665–672, 1958.

[AMA 71] AMAT G., NIELSEN H.H., TARRAGO G., *Rotation-Vibration of Polyatomic Molecules*, M. Dekker Inc., New York, 1971.

[AMA 80a] AMAT G., Cours de Physique Moléculaire, Maitrise de Physique 2e année, University Course, UPMC, 1980.

[AMA 80b] AMAT G., Spectroscopie des Constituants Atmosphériques, Maitrise de Physique 2e année, University Course, UPMC, 1980.

[AND 49] ANDERSON P.W., "Pressure Broadening in the Microwave and Infra-Red Regions", *Phys. Rev.*, vol. 76, pp. 647–661, 1949.

[ANT 06] ANTONY B.K., GAMACHE P.R., SZEMBEK C.D. *et al.*, "Modified complex Robert-Bonamy formalism calculations for strong to weak interacting systems", *Mol. Phys.*, vol. 104, pp. 1–9, 2006.

[BAR 58] BARANGER M., "General Impact Theory of Pressure Broadening", *Phys. Rev.*, vol. 112, pp. 855–865, 1958.

[BAR 61] BARCHEWITZ P., *Spectroscopie infrarouge 1: Vibrations moléculaires*, Gauthier-Villars, Paris, 1961.

[BAR 62] BARANGER M., "Spectral Line Broadening in Plasmas", in BATES D. (ed.), *Atomic and Molecular Processes*, Academic Press, New York, 1962.

[BEH 02] BEHRENDT A., NAKAMURA T., "Calculation of the calibration constant of polarization lidar and its dependecy on atmospheric temperature", *Optics Express*, vol. 10, pp. 805–817, 2002.

[BEN 66a] BEN-REUVEN A., "Symmetry Considerations in Pressure-Broadening Theory", *Phys. Rev.*, vol. 141, pp. 34–40, 1966.

[BEN 66b] BEN-REUVEN A., "Impact Broadening of Microwave Spectra", *Phys. Rev.*, vol. 145, pp. 7–22, 1966.

[BEN 75] BEN-REUVEN A., *Spectral Line Shapes in Gases in the Binary-Collision Approximation*, in PRIGOGINE I., RICE S.A. (eds), *Advances in Chemical Physics*, vol. 33, John Wiley & Sons, New York, 1975.

[BER 12] BERTAUX J.-L., GONDET B., LEFÈVRE F. *et al.*, "First detection of O2 1.27 μm nightglow emission at Mars with OMEGA/MEX and comparison with general circulation model predictions", *J. Geophys. Res.*, vol. 117, E00J04, pp. 1–9, 2012.

[BLO 58] BLOCH C., "Sur la détermination de l'état fondamental d'un système de particules", *Nuc. Phys.*, vol.7, pp. 451–458, 1958.

[BON 77] BONAMY J., BONAMY L., ROBERT D., "Overlapping effects and motional narrowing in molecular band shapes: Application to the Q branch of HD", *J. Chem. Phys.*, vol. 67, pp. 4441–4453, 1977.

[BOU 82] BOULET C., ROBERT D., "Short time behavior of the dipole autocorrelation function and molecular gases absorption spectrum", *J. Chem. Phys.*, vol. 77, pp. 4288–4299, 1982.

[BRA 95] BRAUN T., DIOT C., HOGLANDER A. *et al.*, An Experimental User Level of Implementation of TCP, PhD Thesis, INRIA, 1995.

[BRO 03] BROWN J.R., CARRINGTON A., *Rotational Spectroscopy of Diatomic Molecules*, Cambridge University Press, Cambridge, 2003.

[BRU 55] BRUCKNER K.A., "Many-Body Problem for Strongly Interacting Particles II: Linked Cluster Expansion", *Phys. Rev.*, vol. 100, pp. 36–45, 1955.

[BUN 98] BUNKER P.R., JENSEN P., *Molecular Symmetry and Spectroscopy*, NRC Research Press, Ottawa, 1998.

[BUT 71] BUTCHER R.J., WILLETTS D.V., JONES W.J., "On the use of a Fabry-Perot etalon for the determination of rotational Raman constants of simple molecules - the pure rotational Raman spectra of oxygen and nitrogen", *Proc. Roy. Soc. Lond.*, vol. 324, no. 1557, pp. 231–245, 1971.

[CAM 85] CAMY-PEYRET C., FLAUD J.M., "Vibration-Rotation Dipole Moment Operator for Asymmetric Rotors", in NARAHARI RAO K. (ed.), *Molecular Spectroscopy: Modern Reseach*, vol. 3, Acadmic Press, Orlando, 1985.

[CAM 88] CAMY-PEYRET C., FLAUD J.M., Cours de Physique Moléculaire, DEA Lasers et Matière, University Course, UPSUD, 1988.

[CAR 70] CARRUTHERS G.C., "Rocket Observation of Interstellar Molecular Hydrogen", *Astrophysical Journal*, vol. 161, pp. L81–L85, 1970.

[CAR 80] CARLOS W.E., COLE M.W., "Interaction between a He atom and a graphite surface", *Surface Science*, vol. 91, pp. 339–357, 1980.

[CLA 75] CLARON HOSKINS L., "Pure rotational Raman spectroscopy of diatomic molecules", *Journal of Chemical Education*, vol. 52, pp. 568–572, 1975.

[COH 73] COHEN-TANNOUDJI C., DIU B., LALOË F., *Mécanique Quantique*, Herman, Paris, 1973.

[CON 79] CONNES P., NOXON J.F., TRAUB W.A. *et al.*, "O_2 ($^1\Delta$) Emission in the Day and Night Airglow of Venus", *Astrophysical Journal*, vol. 233, pp. L29–L32, 1979.

[DAH 88] DAHOO P.R., Sur l'intensité et l'élargissement par la pression des raies de vibration-rotation des bandes ν_3 , $\nu_1+\nu_3$ et $\nu_1+\nu_3-2\nu_2^0$ de $^{12}C^{16}O_2$ et ν_3 de $^{14}N_2{}^{16}O$: Etude expérimentale et interprétation des paramètres d'élargissement au moyen de modèles théoriques semi-classiques, PhD Thesis, UPMC, Paris, 1988.

[DAH 16] DAHOO P.R., POUGNET P., EL HAMI A., *Nanometer-Scale Defect Detection Using Polarized Light*, ISTE, London and John Wiley & Sons, New York, 2016.

[DEM 94] DEMEURE I., FARHAT J., "Systèmes de processus légers: concepts et exemples", *Technique et Science Informatiques*, vol. 13, no. 6, pp. 765–795, 1994.

[DIC 53] DICKE R.H., "The effect of collisions upon the Doppler Width of Spectral lines", *Phys. Rev.*, vol. 89, pp. 472–476, 1953.

[DIC 65] DICKE R.H., PEEBLES P.J.E., ROLL P.G. *et al.*, "Cosmic Black-Body Radiation", *The Astrophysical Journal*, vol. 142, pp. 414–419, 1965.

[DIC 78] DICKMAN R.L., "The Ratio of Carbon Monoxide to Molecular Hydrogen in Interstellar Dark Clouds", *Astrophysical Journal*, vol. 37, pp. 407–427, 1978.

[DOU 41] DOUGLAS A.E., HERZBERG G., "Note on CH^+ in Interstellar Space and in the Laboratory", *Astrophysical Journal*, vol. 94, p. 381, 1941.

[DUB 75] DUBOST H., Étude des mouvements moléculaires, de la relaxation et des transferts d'énergie vibrationnelle de l'oxyde de carbone isolé en matrice de gaz rare à basse température, PhD Thesis, UPSUD, Orsay, 1975.

[DUB 76] DUBOST H., CHARNEAU R., "Infrared absorption spectra of carbon monoxide in rare gas matrices", *Chem. Phys.*, vol. 12, p. 407, 1976.

[DYS 49] DYSON F.J., "The Radiation Theories of Tomonaga, Schwinger, and Feynman", *Phys. Rev.*, vol. 75, pp. 486–502, 1949.

[EDM 57] EDMONDS A.R., *Angular Momentum in Quantum Mechanics*, Princeton University Press, Princeton, 1957.

[ENC 75] ENCRENAZ P.J., FALGARONE E., LUCAS R., "CO, dust and H2 in the molecular cloud near rho Ophiuchi", *Astrophysical Journal*, vol. 44, pp. 73–80, 1975.

[FAN 57] FANO U., "Description of States in Quantum Mechanics by Density Matrix and Operator Techniques", *Rev. of Mod. Phys.*, vol. 29, pp. 74–93, 1957.

[FAN 63] FANO U., "Pressure broadening as a prototype of relaxation", *Phys. Rev.*, vol. 131, pp. 259–268, 1963.

[FEL 79] FELDMAN P.D., MOOS H.W., CLARKE J.T. *et al.*, "Identification of the UV nightglow from Venus", *Nature*, vol. 279, pp. 221–222, 1979.

[FLI 62] FLINN P.A., MARADUDIN A.A., "Distorsion of crystals by point defects", *Annals of Phys.*, vol. 18, pp. 81–109, 1962.

[GAL 61] GALATRY L., "Simultaneous, effect of Doppler and foreign gas broadening on spectral lines", *Phys. Rev.*, vol. 122, pp. 1218–1223, 1961.

[GAM 97] GAMACHE R.R., LYNCH R., PLATEAUX J.J. *et al.*, "Halfwidths and line shifts of water vapor broadened by CO_2: measurements and complex Robert-Bonamy formalism calculations", *J. Quant. Spectr. Rad. Transfer*, vol. 57, pp. 485–496, 1997.

[GAM 98] GAMACHE R.R., LYNCH R., NESHYBA S.P., "New developments in the theory of pressure-broadening and pressure-shifting of spectral lines of H_2O: The complex Robert-Bonamy formalism", *J. Quant. Spectr. Rad. Transfer*, vol. 59, pp. 319–335, 1998.

[GEE 79] GEERLINGS P., BERCKMANS D., FIGEYS H.P., "The Influence of Electrical and Mechanical Anharmonicity on the Vibrational Transition Moments of Diatomic and Polyatomic Molecules", *J. Mol. Struc.*, vol. 57, pp. 283–297, 1979.

[GER 08] GÉRARD J.-C., COX C., SAGLAM A. *et al.*, "Limb observations of the ultraviolet nitric oxide nightglow with SPICAV on board Venus Express", *J. Geophys. Res.*, vol. 113, E00B03, 2008.

[GER 09a] GÉRARD J.-C., COX C., SORET L. *et al.*, "Concurrent observations of the ultraviolet nitric oxide and infrared O2 nightglow emissions with Venus Express", *J. Geophys. Res.*, vol. 114, E00B44, 2009.

[GER 09b] GÉRARD J.-C., SAGLAM A., PICCIONI G. *et al.*, "Atomic oxygen distribution in the Venus mesosphere from observations of O2 infrared airglow by VIRTIS-Venus Express", *Icarus*, vol. 199, pp. 264–272, 2009.

[GER 71] GERSTEN J.I., "Asymptotic Line Shape in Collision-Induced Light Scattering", *Phys. Rev.*, vol. A4, pp. 98–107, 1971.

[GOL 56] GOLDSMITH M., AMAT G., NIELSEN H.H., "Higher Order Rotation-Vibration Energies of Polyatomic Molecules I", *J. Chem. Phys.*, vol. 24, pp. 1178–1182, 1956.

[GOL 57] GOLDSTONE J., "Derivation of the Brueckner Many-Body Theory", *Proc. Roy. Soc.*, vol. A239, pp. 267–279, 1957.

[GON 10] GONDET B., BIBRING J.P., BERTAUX J.L. *et al.*, "O2 nightglow emission detected in Mars atmosphere by OMEGA/MarsExpress", *Comm. European Geosciences Union General Assembly*, Vienna, 3 May 2010.

[GOR 68] GORDON R.G., "Correlation functions for molecular motion", *Adv. Magn. Resonance*, vol. 3, pp. 1–42, 1968.

[GRA 66] GRAY C.G., VAN KRANENDONK J., "Calculation of the Pressure Broadening of Rotational Raman Lines due to Multipolar and Dispersion Interaction", *Can. J. Phys.*, vol. 44, pp. 2411–2430, 1966.

[GRA 68] GRAY C.G., "On the theory of multipole interactions", *Can. J. Phys.*, vol. 46, pp. 135–139, 1968.

[GRA 88] GRAYBEAL J.D., *Molecular Spectroscopy*, McGraw-Hill, New York, 1988.

[HAR 08] HARTMANN J.M., BOULET C., ROBERT R., *Collisional Effects on Molecular Spectra: Laboratory Experiments and Models, Consequences for Applications,* Elsevier B.V., Amsterdam, 2008.

[HAU 80] HAUCHECORNE A., CHANIN M.L., "Density and temperature profiles obtained by lidar between 35 and 70 km", *Geophys. Res. Lett.*, vol. 7, pp. 565–568, 1980.

[HAU 16] HAUCHECORNE A., KECKHUT P., MARISCAL J.-F. *et al.*, "Innovative rotational Raman lidar to measure the temperature profile from the surface to 30 km altitude", *The 27th International Laser Radar Conference (ILRC 27)*, EPJWeb of Conferences, EDP Sciences, vol. 119, p. 06008, 2016.

[HER 50] HERZBERG G., *Molecular Spectra and Molecular Structure: Spectra of Diatomic Molecules,* vol. 1, 2nd ed., D. Van Nostrand, New York, 1950.

[HER 59] HERZFELD K.F., LITOVITZ T.A., Absorption and Dispersion of Ultrasonic Waves, Academic Press, New York, 1959.

[HIR 64] HIRSCHFELDER J.O., CURTISS C.F., BIRD R.B., *Molecular Theory of Gases and Liquids,* Wiley, New York, 1964.

[HOL 86] HOLLAS J.M., *Modern Spectroscopy*, 4th ed., John Wiley & Sons Ltd, 1986.

[JOR 34] JORDAHL O.M., "The Effect of Crystalline Electric Fields on the Paramagnetic Susceptibility of Cupric Salts", *Phys. Rev.*, vol. 45, pp. 87–97, 1934.

[JOR 74a] JORGENSEN F., PEDERSEN T., "A Projector Formulation for the Van Vleck Transformation I: Degenerate case", *Mol. Phys.*, vol. 27, pp. 33–47, 1974.

[JOR 74b] JORGENSEN F., PEDERSEN T., "A Projector Formulation for the Van Vleck Transformation II: Near-degenerate case", *Mol. Phys.*, vol. 27, pp. 959–968, 1974.

[KEC 90] KECKHUT P., CHANIN M.L., HAUCHECORNE A., "Stratospheric temperature measurements using Raman lidar", *Applied Optics.*, vol. 29, pp. 5182–5186, 1990.

[KEM 37] KEMBLE E.C., *The Fundamental Principles of Quantum Mechanics*, Dover Publications Inc., New York, 1937.

[KOL 58] KOLB A.C., GRIEM J., "Theory of Line Broadening in Multiplet Spectra", *Phys. Rev.*, vol. 111, pp. 514–521, 1958.

[KOL 97] KOLSKI C., *Interfaces homme machine*, Hermès, Paris, 1997.

[KON 83] KONO H., LIN S.H., "Spectroscopic properties and relaxation processes of impurity molecules in solids I: Rotational spectra", *J. Chem. Phys.,* vol. 78, pp. 2607–2620, 1983.

[KUB 57] KUBO R., "Statistical-Mechanical Theory of Irreversible Processes I: General Theory and Simple Applications to Magnetic and Conduction Problems", *J. Phys. Soc. Jpn*, vol. 12, pp. 570–586, 1957.

[KUB 62] KUBO R., "Generalized cumulant expansion method", *J. Phys. Soc. Jpn*, vol. 17, pp. 1100–1120, 1962.

[LAL 97] LALLOUET A., "DP-LOG: un langage logique data-parallèle", *Actes des 6 èmes journées francophones de programmation logique et programmation par contraintes JFPLC'97*, pp. 53–68, Orleans, 26–28 May 1997.

[LAN 75] LANDAU L., LIFCHITZ E., *Physique Théorique : Tome 3 Mécanique Quantique*, vol. 3, 3rd ed., Mir, Moscow, 1975.

[LAN 99] LANGLET J., CAILLET J., ALLAVENA M. *et al.*, "Infrared absorption spectra of carbon monoxide in rare gas matrices", *J. Mol. Struct.,* vol. 484, pp. 145–159, 1999.

[LEF 04] LEFEVRE-BRION H., FIELD R.W., *The Spectra and Dynamics of Diatomic Molecules*, Revised and Enlarged Edition, Elsevier, Amsterdam, 2004.

[LIP 50] LIPPMAN B., SCHWINGER J., "Variational Principles for Scattering Processes I", *Phys. Rev.*, vol. 79, pp. 469–480, 1950.

[LYN 96] LYNCH R., GAMACHE R.R., NESHYBA S.P., "Fully complex implementation of the Robert–Bonamy formalism: Half widths and line shifts of H_2O broadened by N_2", *J. Chem. Phys.*, vol. 105, pp. 5711–5721, 1996.

[MA 07] MA Q., TIPPING R.H., BOULET C., "Modification of the Robert-Bonamy formalism in calculating Lorentzian half-widths and shifts", *J. Quant. Spectr. Rad. Transfer*, vol. 103, pp. 588–596, 2007.

[MAC 76] MACRURY T.B., STEELE W.A., BERNE B.J., "Intermolecular potential models for anisotropic molecules, with applications to N_2, CO_2, and benzene", *J. Chem. Phys.*, vol. 64, pp. 1288–1299, 1976.

[MAN 68] MANNHEIM P.D., "Influence of force-constant changes on the lattice dynamics of cubic crystals with point defects", *Phys. Rev.*, vol. 165, pp. 1011–1018, 1968.

[MAN 71] MANNHEIM P.D., COHEN S.S., "Force constant changes and the crystal impurity problem ", *Phys. Rev.*, vol. B4, pp. 3748–3756, 1971.

[MAR 65] MARADUDIN A.A., "The effects of point defects on the vibrations of crystal lattices", *Repts. Prog. Phys.*, vol. 28, pp. 331–380, 1965.

[MCK 40] MCKELLAR A., "Evidence for the Molecular Origin of Some Hitherto Unidentified Interstellar Lines", *Publications of the Astronomical Society of the Pacific*, vol. 52, pp. 187–192, 1940.

[MES 64] MESSIAH A., *Mécanique Quantique*, vols. 1 and 2, Dunod, Paris, 1964.

[MIL 72] MILLS I.M., "Vibration-Rotation Structure in Asymmetric and Symmetric-Top Molecules", in NARAHARI RAO K., WELDON MATHEWS C. (eds), *Molecular Spectroscopy: Modern Reseach*, vol. 1, Academic Press, New York, 1972.

[MUR 80] MURTHY C.S., SINGER K., KLEIN M.L. *et al.*, "Pairwise additive effective potentials for nitrogen", *Mol. Phys.*, vol. 41, pp. 1387–1399, 1980.

[MUR 81] MURTHY C.S., SINGER K., MCDONALD I.R., "Interaction site models for carbon dioxide", *Mol. Phys.*, vol. 44, pp. 135–143, 1981.

[NEL 64] NELKIN M., GHATAK A., "Simple Binary Collision Model for Van Hove's Gs(r, t)", *Phys. Rev.*, vol. 135, pp. A4–A9, 1964.

[NES 94] NESHYBA S.P., LYNCH R., GAMACHE R.R. *et al.*, "Pressure-induced widths and shifts for the ν3 band of methane", *J. Chem. Phys.*, vol. 101, pp. 9412–9421, 1994.

[NIE 51] NIELSEN H.H., "The Vibration-Rotation Energies of Molecules", *Rev. Mod. Phys.*, vol. 23, pp. 90–136, 1951.

[OOB 72] OOBATAKE M., OOI T., "Determination of energy parameters in lennardjones potentials from second virial coefficients", *Prog. Theor. Phys.*, vol. 48, pp. 2132–2143, 1972.

[PAP 82] PAPOUSEK D., ALIEV D., *Molecular Vibration-Rotation Spectra,* Elsevier, Amsterdam, 1982.

[PEN 65] PENZIAS A.A., WILSON R.W., "A Measurement of Excess Antenna Temperature at 4080 Mc/s", *Astrophysical Journal*, vol. 142, pp. 419–421, 1965.

[RAD 95] RADUCU V., JASMIN D., DAHOO R. *et al.*, "Conformational dynamics of the CO:CO2 complex in argon matrices I: Thermodynamical considerations derived from the observed kinetics", *J. Chem.Phys.,* vol. 102, pp. 9235–9239, 1995.

[RAD 96] RADUCU V., GAUTHIER-ROY B., DAHOO R. *et al.*, "The CO:CO2 complex in argon matrices: Experimental evidence for two conformations with spontaneous interconversion", *J. Chem.Phys.*, vol. 105, pp. 10092–10098, 1996.

[RAU 67] RAUTIAN S.G., SOBEL'MAN I.I., "The effect of collisions on the Doppler broadening of spectral lines", *Soviet Physics Uspekh.*, vol. 9, pp. 701–716, 1967.

[ROB 76] ROBERT D., GALATRY L., "Influence of molecular nonrigidity on the infrared absorption and Raman scattering line shape in dense media", *J. Chem. Phys.*, vol. 64, pp. 2721–2734, 1976.

[ROB 79] ROBERT D., BONAMY J., "Short range force effects in semiclassical molecular line broadening calculations", *J. Phys.*, vol. 40, pp. 923–943, 1979.

[ROS 67] ROSE M.E., *Elementary Theory of Angular Momentum*, Wiley, New York, 1967.

[SCH 52] SCHWARTZ R.N., SLAWSKY Z.I., HERZFELD K.F., "Calculation of vibrational relaxation times in gases", *J. Chem. Phys.*, vol. 20, pp. 1591, 1952.

[SIB 88] SIBERT E.L., "Theoretical Studies of Vibrationally Excited Polyatomic Molecules Using Canonical VanVleck Perturbation Theory", *J. Chem. Phys.*, vol. 88, pp. 4378–4390, 1988.

[STE 79] STEWART P., BARTH C.A., "Ultraviolet night airglow of Venus", *Science*, vol. 205, pp. 59–62, 1979.

[SUG 65] SUGDEN T.M., KENNEY C.N., *Microwave Spectroscopy of Gases,* Van Nostrand Company Ltd, London, 1965.

[SUG 99] SUGNY D., JOYEUX M., "On the Application of Canonical Perturbation Theory to Floppy Molecules", *J. Chem. Phys.*, vol. 112, pp. 31–39, 1999.

[SWI 37] SWINGS P., ROSENFELD L., "Considerations regarding interstellar molecules", *Astrophysical Journal.*, vol. 86, pp. 483–486, 1937.

[TEN 03] TENNYSON J., "Molecules in Space", in WILSON S. (ed.), *Handbook of Molecular Physics and Quantum Chemistry*, John Wiley & Sons, Chichester, 2003.

[TIP 70] TIPPING R.H., HERMAN R.M., "The Impact theory for the noble gas pressure-induced HCl vibration-rotation and pure rotation line widths—I", *J. Quant. Spectr. Rad. Transfer*, vol. 10, pp. 881–896, 1970.

[TSA 62] TSAO P.J., CURNUTTE B., "Line–widths of pressure-broadened spectral lines", *J. Quant. Spectr. Rad. Transfer*, vol. 2, pp. 41–91, 1962.

[TUC 76] TUCKER K.D., DICKMAN R.L., ENCRENAZ P.J. *et al.*, "The relation between carbon monoxide emission and visual extinction in cloud L134", *Astrophysical Journal*, vol. 210, pp. 679–683, 1976.

[VAN 29] VAN VLECK J.H., "On Sigma-Type Doubling and Electron Spin in the Spectra of Diatomic Molecules", *Phys. Rev.*, vol. 33, pp. 467–506, 1929.

[VAN 45] VAN VLECK J.H., WEISSKOPF V.F., "On the Shape of Collision-Broadened Lines", *Rev. of Mod. Phys.*, vol. 17, pp. 227–236, 1945.

[VER 92] VERNOV A., STEELE W.A., "The electrostatic field at a graphite surface and its effect on molecule-solid interactions", *Langmuir*, vol. 8, pp. 155–159, 1992.

[VOT 84] VOTER A.F., DOLL J.D., "Transition state theory description of surface self-diffusion: Comparison with classical trajectory results", *J. Chem. Phys.*, vol. 80, pp. 5832–5838, 1984.

[WEI 63] WEINREB S., BARRETT A.H., MEEKS M.L. *et al.*, "Radiospectroscopic observations of OH in interstellar space", *Nature*, vol. 200, pp. 829–831, 1963.

[WIL 70] WILSON R.W., JEFFERTS K.B., PENZIAS A.A., "Carbon monoxide in the Orion Nebula", *Astrophysical Journal*, vol. 161, pp. L43 L44, 1970.

[WIL 80] WILSON E.B., DECIUS J.C., CROSS P.C., *Molecular Vibrations: The Theory of Infrared and Raman Vibrational Spectra*, Revised edition, Dover Publications, New York, 1980.

[YAO 76] YAO S.J., OVEREND J., "Vibrational Intensities XXIII: The Effect of Anharmonicity on the Temperature Dependence of Integrated Band Intensities", *Spectrochimica Acta*, vol. 32A, pp. 1059–1065, 1976.

[ZWA 60] ZWANZIG R., "Ensemble method in the theory of irreversibility", *J. Chem. Phys.*, vol. 33, pp. 1338–1341, 1960.

[ZWE 98] ZWEIFEL H., *Stabilisation of Polymeric Materials*, Springer, Berlin, 1998.

Index

E, F

G, H

I

S

T

U, V, W

Other titles from

ISTE

in

Waves

2017

PARET Dominique, HUON Jean-Paul
Secure Connected Objects

PARET Dominque, SIBONY Serge
Musical Techniques: Frequencies and Harmony

STAEBLER Patrick
Human Exposure to Electromagnetic Fields

2016

ANSELMET Fabien, MATTEI Pierre-Olivier
Acoustics, Aeroacoustics and Vibrations

BAUDRAND Henri, TITAOUINE Mohammed, RAVEU Nathalie
The Wave Concept in Electromagnetism and Circuits: Theory and Applications

PARET Dominique
Antennas Designs for NFC Devices

PARET Dominique
Design Constraints for NFC Devices

WIART Joe
Radio-Frequency Human Exposure Assessment

2015

PICART Pascal
New Techniques in Digital Holography

2014

APPRIOU Alain
Uncertainty Theories and Multisensor Data Fusion

JARRY Pierre, BENEAT Jacques N.
RF and Microwave Electromagnetism

LAHEURTE Jean-Marc
UHF RFID Technologies for Identification and Traceability

SAVAUX Vincent, LOUËT Yves
MMSE-based Algorithm for Joint Signal Detection, Channel and Noise Variance Estimation for OFDM Systems

THOMAS Jean-Hugh, YAAKOUBI Nourdin
New Sensors and Processing Chain

TING Michael
Molecular Imaging in Nano MRI

VALIÈRE Jean-Christophe
Acoustic Particle Velocity Measurements using Laser: Principles, Signal Processing and Applications

VANBÉSIEN Olivier, CENTENO Emmanuel
Dispersion Engineering for Integrated Nanophotonics

2013

BENMAMMAR Badr, AMRAOUI Asma
Radio Resource Allocation and Dynamic Spectrum Access

BOURLIER Christophe, PINEL Nicolas, KUBICKÉ Gildas
Method of Moments for 2D Scattering Problems: Basic Concepts and Applications

GOURE Jean-Pierre
Optics in Instruments: Applications in Biology and Medicine

LAZAROV Andon, KOSTADINOV Todor Pavlov
Bistatic SAR/GISAR/FISAR Theory Algorithms and Program Implementation

LHEURETTE Eric
Metamaterials and Wave Control

PINEL Nicolas, BOURLIER Christophe
Electromagnetic Wave Scattering from Random Rough Surfaces: Asymptotic Models

SHINOHARA Naoki
Wireless Power Transfer via Radiowaves

TERRE Michel, PISCHELLA Mylène, VIVIER Emmanuelle
Wireless Telecommunication Systems

2012

LALAUZE René
Chemical Sensors and Biosensors

LE MENN Marc
Instrumentation and Metrology in Oceanography

LI Jun-chang, PICART Pascal
Digital Holography

2011

BECHERRAWY Tamer
Mechanical and Electromagnetic Vibrations and Waves

GOURE Jean-Pierre
Optics in Instruments

GROUS Ammar
Applied Metrology for Manufacturing Engineering

LE CHEVALIER François, LESSELIER Dominique, STARAJ Robert
Non-standard Antennas

2010

BEGAUD Xavier
Ultra Wide Band Antennas

MARAGE Jean-Paul, MORI Yvon
Sonar and Underwater Acoustics

2009

BOUDRIOUA Azzedine
Photonic Waveguides

BRUNEAU Michel, POTEL Catherine
Materials and Acoustics Handbook

DE FORNEL Frederique, FAVENNEC Pierre-Noël
Measurements using Optic and RF Waves

FRENCH COLLEGE OF METROLOGY
Transverse Disciplines in Metrology

2008

FILIPPI Paul J.T.
Vibrations and Acoustic Radiation of Thin Structures

LALAUZE René
Physical Chemistry of Solid-Gas Interfaces

2007

KUNDU Tribikram
Advanced Ultrasonic Methods for Material and Structure Inspection

PLACKO Dominique
Fundamentals of Instrumentation and Measurement

RIPKA Pavel, TIPEK Alois
Modern Sensors Handbook

2006

BALAGEAS Daniel *et al.*
Structural Health Monitoring

BOUCHET Olivier *et al.*
Free-Space Optics

BRUNEAU Michel, SCELO Thomas
Fundamentals of Acoustics

FRENCH COLLEGE OF METROLOGY
Metrology in Industry

GUILLAUME Philippe
Music and Acoustics

GUYADER Jean-Louis
Vibration in Continuous Media

Printed and bound by CPI Group (UK) Ltd, Croydon, CR0 4YY